William Winthrop Kent

Architectural wrought-iron, ancient and modern

A compilation of examples from various sources of German, Swiss, Italian,

French, English and American iron-work from mediaeval times down to

the present day

William Winthrop Kent

Architectural wrought-iron, ancient and modern
*A compilation of examples from various sources of German, Swiss, Italian, French,
English and American iron-work from mediaeval times down to the present day*

ISBN/EAN: 9783337229061

Printed in Europe, USA, Canada, Australia, Japan

Cover: Foto ©berggeist007 / pixelio.de

More available books at **www.hansebooks.com**

ARCHITECTURAL

WROUGHT-IRON

ANCIENT AND MODERN.

A COMPILATION OF EXAMPLES FROM VARIOUS SOURCES, OF GERMAN,
SWISS, ITALIAN, FRENCH, ENGLISH AND AMERICAN IRON-
WORK FROM MEDIÆVAL TIMES DOWN TO
THE PRESENT DAY.

BY

WILLIAM WINTHROP KENT,

ARCHITECT

NEW YORK.

WM. T. COMSTOCK,

23 WARREN STREET.

1888.

DEDICATION.

To the younger members of the profession, with the hope that it will prove of interest and assistance this collection is sincerely dedicated.

Guard:
Hotel
Clvny

CONTENTS.

PLATES.

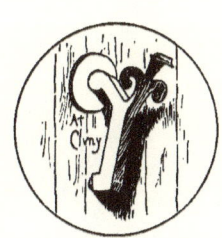

ABBREVIATIONS. Where subjects have been taken from periodicals, the following abbreviations have been used to indicate the source: A, from "English Architectural Association Sketch-Book;" B, "Aus der Kunst-Schmiede-Eisen-Sammlung;" C, English "Building News;" D, "British Architect;" E, Photograph.

PREFACE.

It is only within a comparatively few years that much attention
has been given by American architects to the detail of wrought-iron
design. The causes of this are, I think, the same that have retarded
the progress of architecture and the allied arts generally—want of direct
tradition, and the carelessness natural to a new and money-making
nation, about anything except that which has a practical bearing on
immediate material gain.

That a great change has taken place in this respect, and is exerting
a stronger influence every day on the quality of architectural design
generally, must be evident to anyone who is sufficiently interested in
such matters to look around him. What has caused this change beyond
the sudden acquisition of wealth, and consequent improvement in the
public taste, it is not possible to discuss here, but that bad design should
henceforth be more a matter of stupidity than of custom, is certainly to
be desired and this is, I hope, an excuse for even such modest attempts
as the present compilation.

I wish to acknowledge my indebtedness to Mr. H. C. Burdett,
Architect, for the assistance which several of his excellent sketches have
given, and likewise to Messrs. G. Krug & Son, of Baltimore, and
Mr. Jno. Williams, of New York, for the loan of photographs and
specimens of their iron work.

New York, Sept. 1st, 1888.

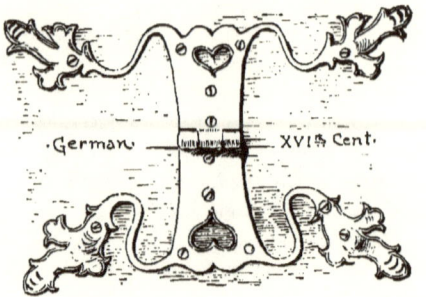

·German· ——— XVIth Cent·

INTRODUCTION.

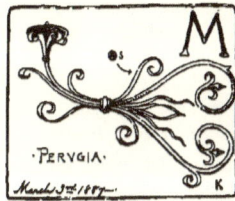

MAN in his inventions does not progress far in the direction of the practical and useful before the spirit of beautifying takes possession of him, and that which at first served him for material ends alone, as it becomes more and more the object of his thought and inspiration, assumes, under his eye and hand, new beauty, until its ornamental character, added to its useful property, makes a more nearly perfect whole, and the craft of the artisan is gradually lost sight of in our admiration of the skill of the developing artist. To use a simple illustration, primeval man made his pottery to serve for purely useful purposes ; then, dissatisfied with its rude appearance as time went on, he decorated it and improved its forms through successive ages, until, in the Greek vase, the potter's art came as near perfection as it has ever come.

I do not mean that that which is intended for purely practical ends is not beautiful, for we know that necessity and the attendant conditions of life often mold the works of man into shapes which all the efforts of would-be-greater descendants fail to improve upon. We see this in the canoe paddle of the Fiji, in the outlines of the bungalow, in the modeling of the Viking ship, in the curve of the hunting bow and the head of the spear. Indeed, it is the exception to find that the inventions born of necessity are deformed offspring, and where man, independent of tradition and precedent, has solved the problems of construction and ornament, there is often a more satisfactory simplicity in the result than is attained by later generations, who follow too closely on the heels of their ancestors. Yet there can be no question that in all the arts and sciences man improves chiefly by retrospective study and research. The workman, whether artist or artisan, works to better advantage if he knows what has been done before—how his brother workman of centuries ago studied and solved problems similar to the ones which arise from the conditions of life to-day.

The malleability and ductility of iron together with its strength, have, ever since man learned its use, made it one of the most serviceable of metals. For weapons of war and the chase, for agricultural implements, and in all its constructive uses, iron, and especially wrought-iron, has been of inestimable value to him, which accounts for the fact that in shaping it for its various offices he has taxed his inventive ability to its utmost limit. Still, out of the old forms spring new ones, better suited to modern needs, and to-day we find it more than ever used and filling a place for which, all things considered, it would be difficult to find a satisfactory substitute. But it is chiefly in connection with architecture that wrought-iron has been used in its most varied and interesting forms, both as regards art and science. Hence, while some small space must necessarily be given to a brief description of the domestic uses to which it has been put, yet in the main this compilation will deal with it in its architectural field, and especially in its decorative use therein.

As one of the baser metals, it is perhaps not strange that among some of the ruder nations such as the mediæval Germans, Norwegians, and Swiss, iron was worked in shapes and patterns more closely in keeping with its peculiar properties than has been the case among later and more cultivated races. Moreover, during the Middle Ages generally, its properties and use seem to have been better understood than they are now. In those days its forms were generally such as suggested its combined baseness, strength and pliability, and in each of the best examples of the old work we usually find all of these characteristics respected. That is to say, we find it first of all in its proper place, serving the purpose for which it was made, and not overwrought or made into forms and patterns which suggest and belong to the more precious metals. To-day we often expend on it all the skill and labor of the goldsmith, shaping it into forms which are not characteristically appropriate, and placing it where, instead of interesting, it repels us. An iron grille is a sensible protection against thieves, and ornamentation does not destroy its usefulness; but when we make iron picture frames, with realistic leaves and flowers, and other similar designs, we go beyond its province.

It is a question how far realism should be followed in any design, but certainly if there is any material which in its decorative treatment demands a close following of conventionalized ornament, that material is wrought-iron. A proof of this is the fact that in what are commonly acknowledged the better examples of the work of respective nations—that is, the examples in which the characteristics of the people and their times are most forcibly expressed—we find little divergence from the path of conventional ornament, and wherever we do find a departure, from that very point we can trace the beginning of a national decline in wrought-iron design. Let us cite one instance. The mediæval Germans, in ornamenting their work with conventional suggestions of the gnomes, goblins and monsters in which their legends abound, imbued their designs with the unmistakable spirit of art, but when they tried to follow nature too closely in the forms of leaf, twig and flower, the failure in design was signal and the descent to mediocrity easy. It may be that this departure from conventional forms is always a sign that the strength of national imagination is growing less, and perhaps in time it must come in all designs. However that may be, it is seen to be universally the case that when the designer ceases to make full use of his imaginative powers, his work ceases to improve and finally deteriorates.

It seems as if there was in the straightforwardness of the mediæval mind a certain sympathy with the properties of iron that led to a wise use of it ; and this might be said of other materials also. In our own day the village blacksmith, probably because he handles iron constantly, is better able to forge a beautiful hinge or finial than half the skilled workmen of the towns, because by constantly using iron he comes to know its properties, and so can put his spirit into it, and is not afraid of the marks of his hammer. This is why, in his work, we find all the strength and crispness of an artist's first sketch.

The best ancient work in iron seems to have been done from the tenth to the seventeenth century. At least we find it was in these centuries that most of the best now remaining examples were executed. Italy, France, Germany and Switzerland were then notably filled with works, the quality of which the workman of to-day cannot hope to rival until he is better able to appreciate its excellence by thoroughly understanding the spirit which produced it.

That study of the older forms and comparison with the best modern work can in some measure assist to such an understanding, the writer believes, and hence into this collection has been put what seemed best calculated to help the designer and the artisan to a better appreciation of what has constituted and still constitutes good design in wrought-iron.

CHAPTER I.

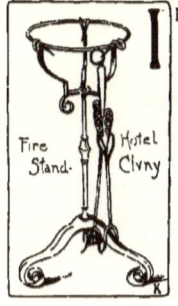

Fire Stand. Hotel Clvny.

I N earlier times, when the state of society was in a more unsettled condition than that which we enjoy to-day, the iron guard in various forms was an important and, in fact, necessary accessory to many buildings. The portcullis, with its sharpened spikes, the window grille, with threatening finials, of which M. Viollet-le-Duc gives several good examples in his *Dictionnaire*, and other less defensive designs, are familiar to all. To-day, the most common form is the window grille, which we use chiefly as a protection in exposed openings, and as these often come on the street, it is important in a house of any pretentions to pay some attention to their design when so located.

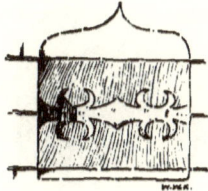

WINDOW GUARD. DIEPPE. FRANCE.
No 1.

Nos. 1 and 2 are rude forms of window guards from old French houses, and being made from one piece, the decorative effect is obtained with little expense, the iron bar being simply cut and branched. By repeating the bar of No. 1 and placing several side by side with the branchings at different distances, very simple and effective guards are obtained, as is shown in No. 2.

No. 3 is a shutter guard, common to many of the Swiss houses at
Amsteg, Altorf, and along the road from Lake Lucerne to the Furca pass.
These are interesting chiefly on account of their primitive simplicity and
because they have also the qualities of originality and strength so often
found in rude work and so often wanting in more elaborate designs.

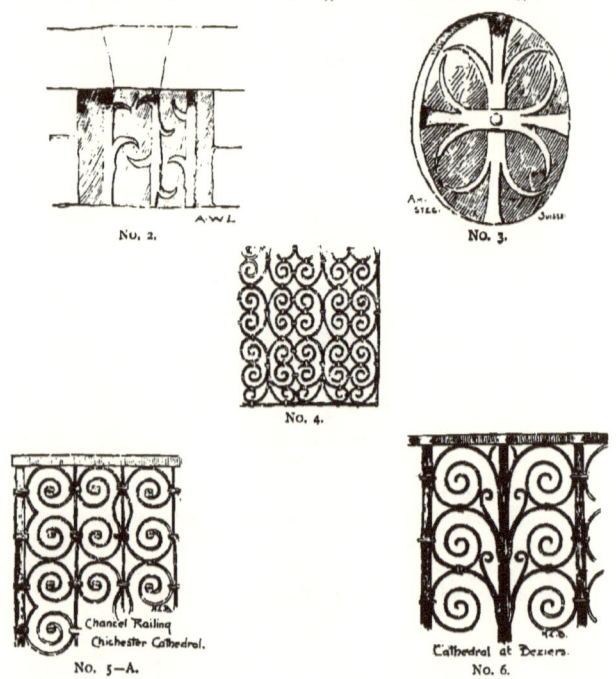

No. 2.

No. 3.

No. 4.

Chancel Railing
Chichester Cathedral.
No. 5—A.

Cathedral at Beziers.
No. 6.

Nos. 4, 5, 6 are examples of very common motives in both ancient and
modern designs.   Shapes like these are often nowadays beaten out in molds
to save the expense of skilled labor.   It is needless to say that such treat

ment robs them at once of all the interest which attaches to hand work and
reduces them to the commonplace level of machine stamping—a fact which
many modern manufacturers are constantly overlooking. Indeed, if iron is
to be worked in molds, why not be honest and make use of cast-iron

No. 7—C.

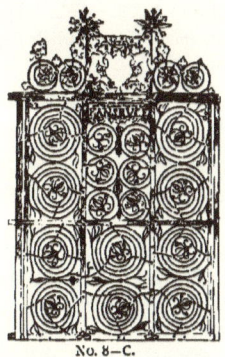

No. 8—C.

whereby the mold fills its proper place? The attempt to cheat the mind
by deceiving the eye in such a way is as short-sighted as it is generally
unsuccessful.

No. 9—B.

No. 7 is a sixteenth century grille of German workmanship of a type
which deserves considerable study. It is simple and relies for effect on the
arrangement of the curves and angles made by a round rod of iron without
resort to any foliation, save a simple separation of the ends.

No. 8, a sixteenth century gate, is one of much the same general character, but in this notice the introduction of the graceful leaf and bud form in the center of the whirls. This piece is dated 1576.

No. 9 is also a German grille, from an excellent book on iron-work, entitled " Aus der Kunst-Schmiede-Eisen-Sammlung des Architekten Friedr. Halselman," and the whole design, of which this is a part, is one of the most refined and beautiful examples of wrought-iron. The splendid conventional

No. 10—B.

treatment of the leaf form is particularly interesting. The rosettes on the bars are perhaps a little stiff, but altogether it is an unusually good grille. Several examples have been taken from this book, which is one of the best collections ever published.

No. 10 is half of a transom grille, or possibly the ornamental top of a

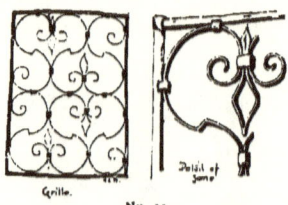

Grille.

Detail of Same

No. 11.

gate or fence. This is also German. The ornament in the lower left-hand corner has a curious resemblance to late Renaissance design.

The detail of No. 11 shows how valuable it is to study in perspective. By making the hoops flat and presenting the narrow edge first to the eye, delicacy and strength of construction are obtained and a constantly varying profile, as seen from different points of view. Whereas, in No. 4, a certain unrelieved flatness is the prevailing characteristic, and the design depends almost wholly for its beauty on the outline, seen from directly in front, while the construction is more difficult and less stable.

In continuing this subject, it will be noticeable that while the ancient work is interesting in many cases, on account of a roughness which is the

CHICHESTER CATH·
No. 12—A.

No. 13—A.

result of hand workmanship, this quality is never imitated, as is often done to-day by false hammer marks, etc. Such an imitation never deceives, and is exceedingly amateurish and weak, like any similar trickery.

No. 12 is a grille which combines delicacy, symmetry, strength and variety of movement to a remarkable degree, and altogether is an excellent design.

No. 14—A.

Venice·Apr·2ᵈ 87·
No. 15.

Nos. 13 and 14 are circular and semi-circular grilles for a round window and a transom light respectively. They are, I believe, Venetian.

Nos. 15 and 16 are also outlines of grilles at Venice—No. 16 being before a shrine of the Madonna; the sketch shows only one-quarter of the design.

No. 17 is a comparatively modern railing from a New York door-step. It shows into what a respectable design even a plain bar of iron without taper can be made.

No. 18 is a French grille, showing an interesting treatment of whirls at

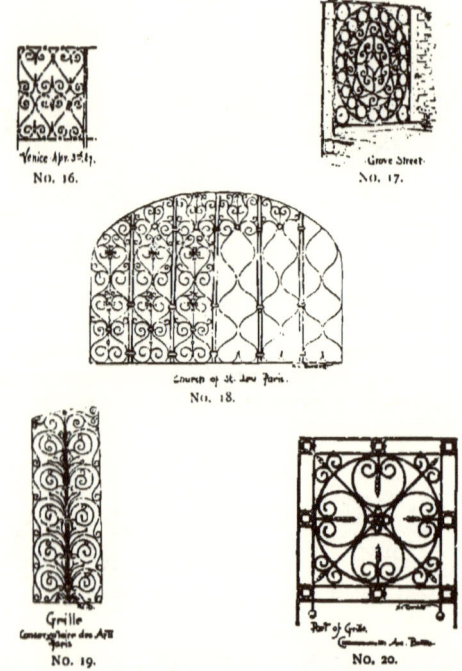

Venice Apr. 3ᵈ. 17.
No. 16.

Grove Street.
No. 17.

Church of St. Leu Paris.
No. 18.

Grille
Conservatoire des Arts
Paris
No. 19.

Part of Grille
Conservatoire des Arts Paris
No. 20.

the top. The cross is also well introduced, without being repeated too often.

No. 19 is a modern French grille with a particularly good ending for the stems at the sides.

No. 20 is a modern American grille which, having the elements of a good design, lacks delicacy, all the bars being monotonously alike and too heavy. Compare this with No. 8 and its faults are easily seen.

from T. L. Higginson's
house Boston.
No. 21.

from The Equitable
Building N. Y.
No. 22.

Nos. 21 and 22 are modern American grilles of considerable merit. No. 21 is from the studio of the late H. H. Richardson, as is also No. 23, which I believe is an unexecuted design.

· Grille ·
No. 24.

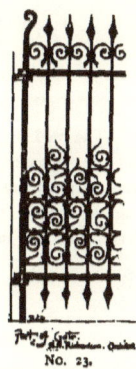

Part of Gate.
H. H. Richardson. Architect.
No. 23.

No. 24 is a modern American design which, though having a plain geometrical frame work, is nevertheless made very graceful and interesting by the skillful disposition of its more delicate members.

CHAPTER II.

ONE of the most important articles of metal-work in public and private use during the Middle Ages was the hinge, which was very generally made of iron, owing to the low cost of this metal and the ease of working it. In household furniture, the chest, cabinet, shutter and door were all furnished with hinges of more or less elaborate design, and among public edifices, the entrance doors both of civil and ecclesiastical buildings, were generally ornamented with strap hinges, which served the double purpose of decorating and binding the door more firmly together, and this last was no small *desideratum* in an age when the changes of government were often sudden and attended with violence.

Sketch of Iron Casket

No. 1—C.

The mediæval workman gave great scope to his imagination in supplying the demand which was caused by such a general use. The Germans were pre-eminently the great hinge-makers, and, as has been before remarked, their best work was unmistakably stamped with characteristic decoration, and in the flatter forms of metal-work, such as hinges, escut-

cheons, etc., they found ample opportunity for the conventional expression
of the weird and peculiar ideas with which not only their folk lore, but even
their minds and literature, were filled. Why they brought the design of
the hinge to such a state of excellence, I have never completely understood,
but that they did must be evident to anyone who has studied carefully the
relative metal-work of the different nations at this period.

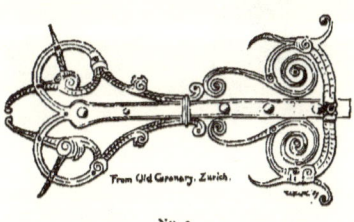

From Old Granary, Zurich.

No. 2.

It is a very singular and interesting fact, this general desire of the
Northern nations to embody in art the spirit which pervaded, and which
still, in a measure, does pervade, their lives and traditions ; but the Teutons

No. 3—B.

gave greater expression to this desire than other races, and in consequence
their work possesses a strong and fascinating individuality.

No. 1 is of German work, probably of the latter part of the fifteenth
or early sixteenth century, and illustrates what has been said of the prev-
alence of grotesque animal forms.   The work is crude in some respects,

but the chisel cuts most effectively atone for the apparent rudeness of outline.

No. 2 is one of a set of twelve hinges of different design on the doors
of the old granary at Zurich, which is situated close by the edge of the
outlet of Lake Zurich. They are full of spirit and thoroughly like the

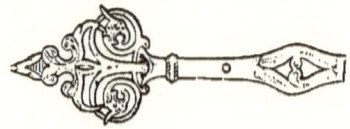

No. 4 –B.

German work in every particular. The sudden shooting out of the two
pieces at the end is in very effective contrast to the undulating curves of
the rest of the strap. Sketches of some of the other hinges are given
among the plates.

Nos. 3 and 4 are German hinges on which the chisel ornaments are

No. 5 B.

particularly good. The scale ornament, made with a gouge, is very
common in early German work.

In No. 4 the bird's head is that of a parrot, and at the other end of the
strap are representations of feathers.

No. 5 is a very delicate hinge, unusually refined for German work. It
is extremely simple, but very graceful. The ends look like Dutch workmanship.

No. 6 is a Gothic design, probably of German make. It has excellent outlines, and the introduction of the leaves into the trefoil is very happily done.

No. 7 is an English door-hinge of about 1600. The edges are chamfered like much of the German work, and the heads at the end are also singularly like the latter school.

No. 8 is a Swiss hinge of the shutter of a house at Zurich. The end is rather refined. It is probably seventeenth century work.

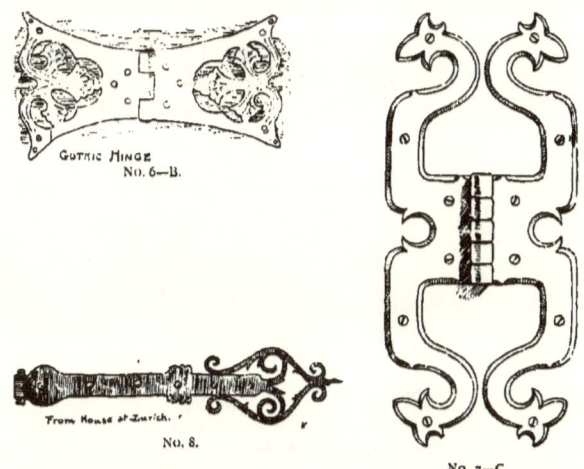

GOTHIC HINGE
No. 6—B.

From House of Zurich.
No. 8.

No. 7—C.

No. 9 is also Swiss, and its curves are remarkably good. There is a large monastery at Einsiedeln, where the working of iron by the brotherhood for the cathedral has had, no doubt, some influence on local designs, although there are very few examples to be seen in the town outside of the church.

No. 10 is a more clumsy one than the preceding, but like it in general outline. A comparison of the two shows wherein the excellence of No. 9 consists.

No. 11 is from a church at Piacenza, Italy, and although apparently the

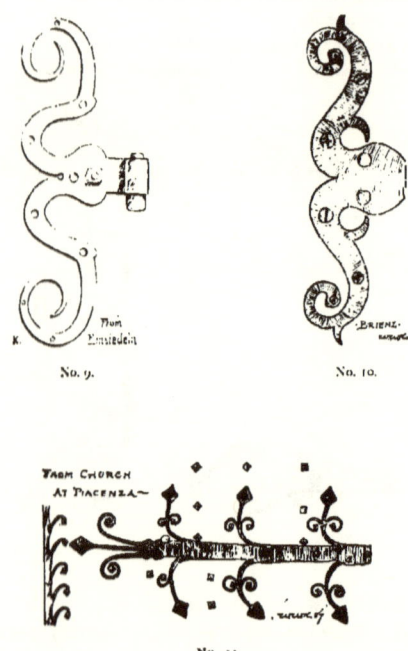

No. 9.   No. 10.

No. 11.

rude work of a blacksmith, it has considerable vigor and the motive of a
very good design.

No. 12 is a modern design on old French lines and its prototype may
be seen on many a provincial church door.

No. 13 is a sketch of hinges for a stable door, serving also to swing

No. 12.

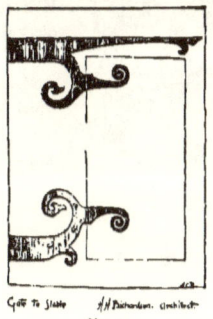

Gate To Stable   *H.H. Richardson. Architect.*
No. 13.

the smaller entrance door cut in it. It expresses decidedly an ancient feel-
ing and I have made use of it as a motive in designing the cover to this
collection.

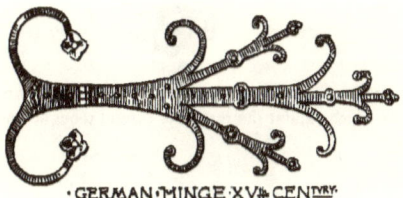

·GERMAN·HINGE·XVᵗʰ·CENTᴜʀʏ·

CHAPTER III.

DOOR KNOCKERS AND RINGS.

THE introduction of bells has almost put an end to the use of knockers, relegating them to country houses and artists' studios, and it is to be regretted, since the modern bell-pull does not give that chance for design which was afforded by the ancient knocker. But worse than this, the electric bell has left only the escutcheon, and a small one at that, on which the imagination of the designer can work. However, even though the knocker

NO. 1.

itself can rarely now-a-days be directly used, there are so many of its old forms which exhibit the results of careful thought and study, that at number of examples are given here which may prove of assistance in other directions.

I have seen very few examples of so-called Colonial knockers, which were of wrought-iron, and so, unfortunately, am unable to give any examples, if indeed any good ones exist. Most of them are made of cast iron, brass or bronze metal, and some of the common forms are very good in design, though usually extremely simple.

As to door rings, there seems to be no good reason why their use to-day should not be to some extent continued. Certainly it is much safer to close a heavy door by a central ring than by the usual knob, since with the

No. 2.

former one runs no risk of shutting in fingers. On church doors and the doors of public buildings it would seem especially advisable to use them more generally than we do, both for the reason above given and for their decorative qualities, which, when combined with those of hinge straps, are considerable.

No. 1 is a rather rude knocker of Spanish workmanship, on which there are several good decorative patterns which, from their character (being easily made with gouges and chisel) are well adapted to the material.

No. 2, a Pisan knocker, has a very graceful ring and an excellent conventional animal head. The entire modelling is extremely delicate and beautiful.

No. 3, which also serves to raise the latch, is a remarkably good design, and the curves are full of the strength and spirit which is so often seen in the French iron-work, and noticeably lacking in a great many of the English designs.

No. 4, which I believe is of Scotch workmanship, is bold in outline, and although perhaps heavy and set in effect, is altogether decorative.

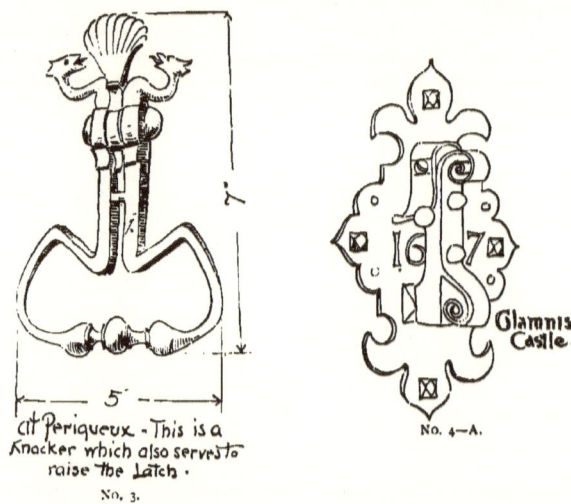

At Periqueux - This is a
Knocker which also serves to
raise the Latch.
No. 3.

Glamnis Castle.

No. 4—A.

No. 5 is a curious piece of Spanish work, suggestive of a forcible entrance. There is nothing valuable about it except the idea, which might be successfully used in a different form.

Nos. 6 and 7 are also Spanish, and show good conventional treatment of animal forms. They have a great deal of rude life and vigor, and yet do not follow nature too closely.

No. 8, from the hospital at Beaune, France, has a fine head well brought out of the rest of the design. There is, I believe, much more good iron work in this building, especially on a Gothic knocker, of which Rague-net gives a cut in the "Documents."

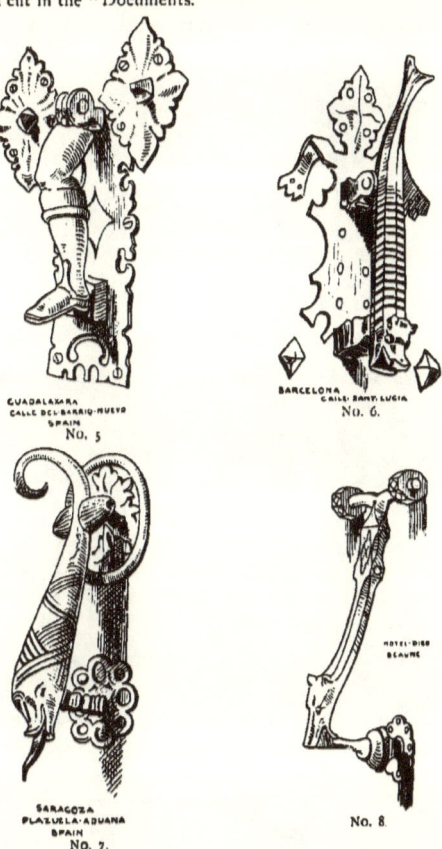

GUADALAXARA
CALLE DEL BARRIO-NUEVO
SPAIN
NO. 5

BARCELONA
CALLE-SANTA LUCIA
No. 6.

SARAGOZA
PLAZUELA-ADUANA
SPAIN
NO. 7.

HOTEL-DIEU
BEAUNE

No. 8.

No. 9 is an extremely simple design, and from its character, I judge the original was a piece of French work, executed under Gothic influence.

No. 10 has nothing but its simplicity and inexpensiveness to recommend it.

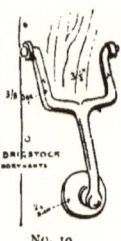

No. 9.   No. 10.

No. 11 shows but little attempt at design, and to judge from its hanging, might not always strike in the right place.

No. 11.   No. 12.

No. 12 has a pretty scutcheon, and with a little study might be much improved.

CHAPTER IV.

ITALIAN WALL-RINGS AND STAPLES.*

Sᴛ Michele— ...
·Pavia·

THESE rings and staples are from the walls of houses and palaces at Siena. They were used in many of the Northern Italian cities during the middle ages, but those we find at Siena and Florence are especially fine examples of decorated iron-work. They were used for various purposes. In times of civil disturbance chains were stretched across the streets to impede the advance of mobs or soldiery. In Florence and Siena during the struggles of the Guelphs and Ghibellines it is said prisoners were chained to them, exposed to the taunts and insults of the people.

As civilization advanced they were used on new buildings more as architectural accessories than from any real need, and to-day they are cherished as much for historical association as for any other reason.

The decorative patterns upon them are very plain, but well suited to the character of the material, while the expression and life-like vigor of the heads is more striking in reality than can be easily indicated in a sketch.

On some buildings they are placed so high above the street between the second-story windows that they must also have served as supports for banners and flags during public festivals, or to hold the ends of awning poles in hot weather.

As architectural accessories and as strong though rude specimens of the smith's art they are interesting reminders of the spirit of one of the most remarkable periods in the history of Northern Italy.

* See Plates No. XX, and XXI.

FINISHES FOR PRESERVING WROUGHT IRON.

There are many ways of finishing iron so as to preserve it from the action of the weather, moisture, etc., but very few of these ways can be recommended.

If paint be used, it is hardly necessary to advise the use of black paint in preference to any color for general work.

A finish which is used by some of the American manufacturers, and said to be good, is the following :

A mixture of asphaltum and lamp black of about the consistency of ordinary varnish applied with a brush and polished when dry. If a lustre is desired a shellac sizing may be added. Black shellac is better than white for this purpose.

This has to be applied two or three times a year on out-of-door work, according to the dampness of the climate.

One other method is the Bower-Barff process, which is good but expensive and not a perfect protection. The process is a secret, but consists apparently in depriving the iron of its rusting qualities by the action of fire or chemicals, or of both. It has the virtue of retaining the original surface of the iron intact.

Iron may also be galvanized and stained with acid stains and then shellaced. This is said to be very durable.

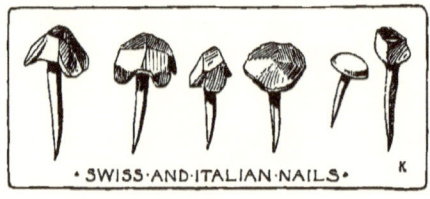

· SWISS·AND·ITALIAN·NAILS·   K

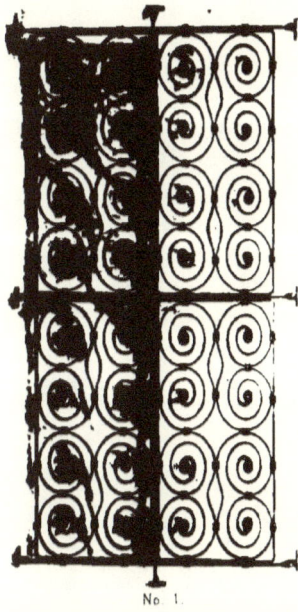

No. 1.

Window Grilles
from
Washington and Baltimore
Houses

No. 1. MESSRS HORNBLOWER AND MARSHALL, ARCHITECTS.

No. 2. IN HOUSE OF MR JOHN HAY, WASHINGTON, D. C. MR H. H. RICHARDSON ARCHITECT.

No. 3. MESSRS McKIM, MEAD & WHITE, ARCHITECTS.

No. 4. IN HOUSE OF MR B. H. WARDER WASHINGTON, D. C. MR H. H. RICHARDSON, ARCHITECT.

No. 2

No. 3.

No. 4

PLAT

·NVREMBVRG·XVIᵗʰ·CENTVRY·

·NVREMBVRG·RATHAVS·1616-19·

·GRILLES·

PLATE II.

ITALIAN
16th Century
South
Kensington.

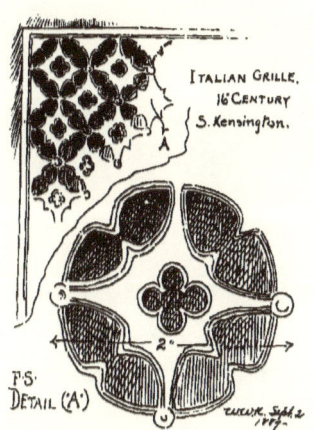

ITALIAN GRILLE.
16 CENTURY
S. Kensington.

F.S.
DETAIL ('A')

WINDOW GRILLES ETC.

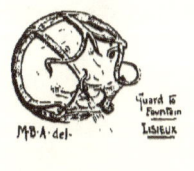

Guard to Fountain
LISIEUX

M.B.A. del.

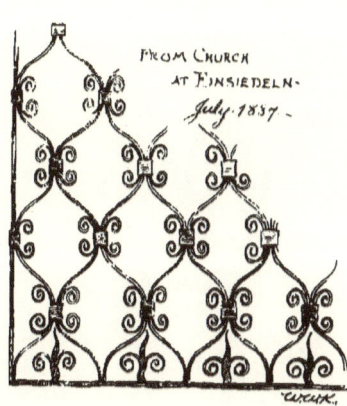

FROM CHURCH
AT EINSIEDELN.
July. 1837.

Venetian Window Grilles

Venetian Grille

ITALIAN   17th CENT-
URY.

PLATE III.

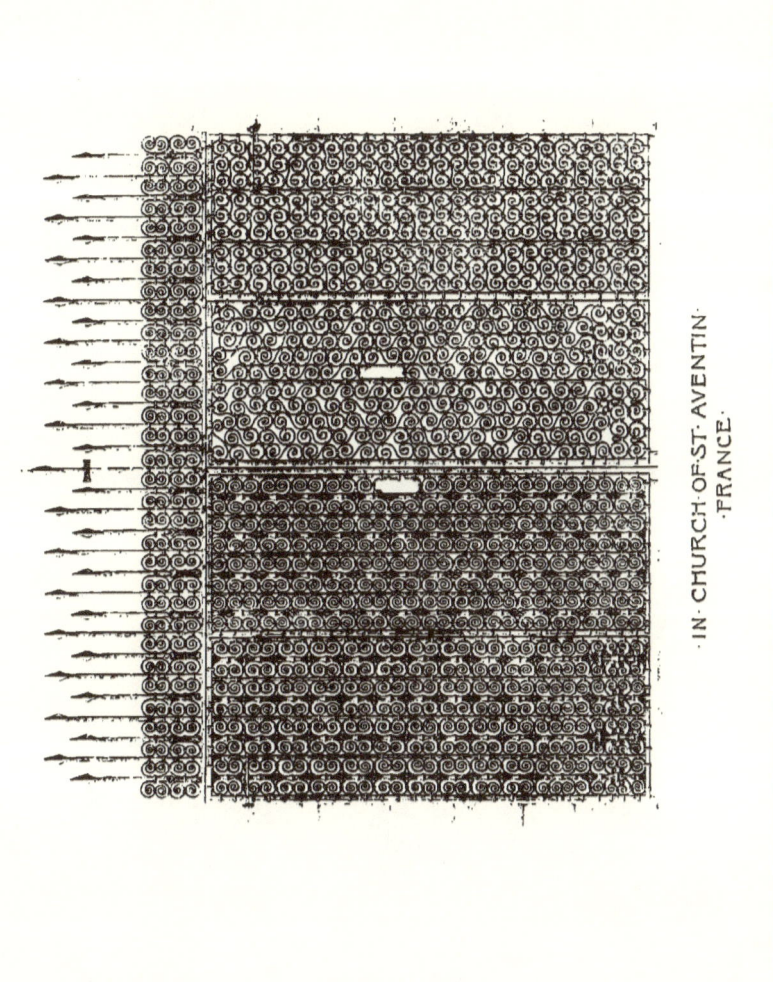

·IN·CHURCH·OF·ST·AVENTIN·
·FRANCE·

PLATE IV.

·FIRE-SCREEN·

C

PLATE V.

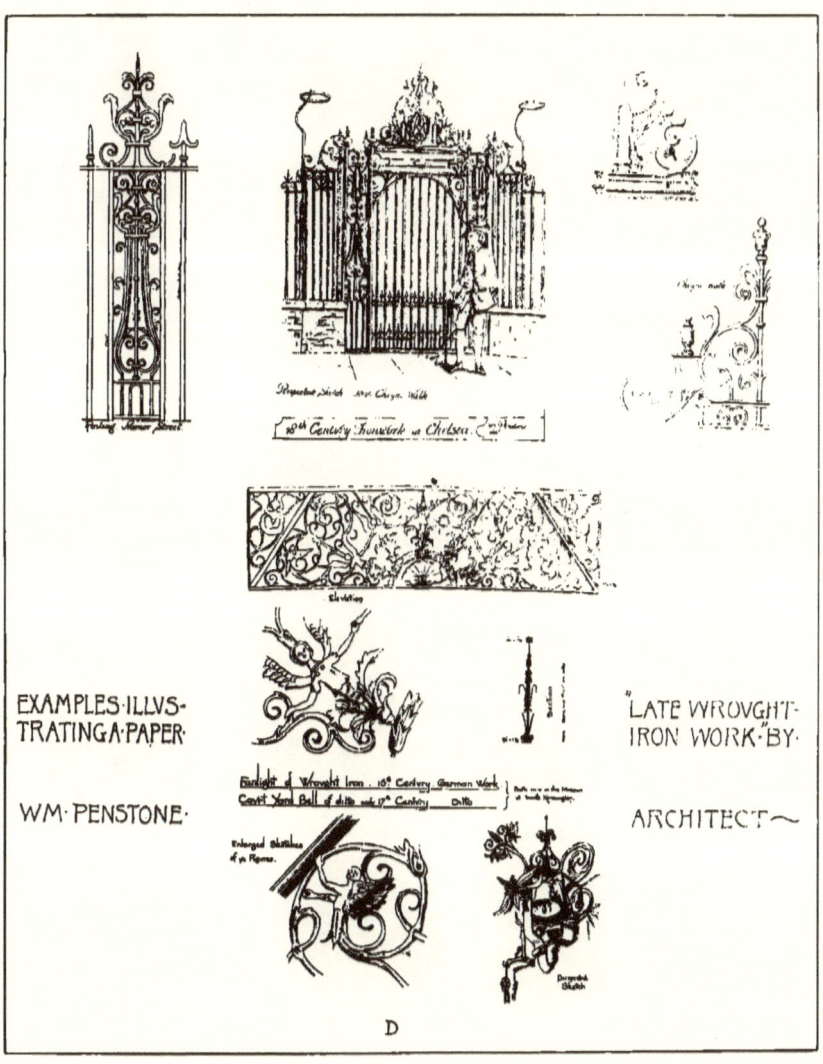

EXAMPLES·ILLVS-
TRATING·A·PAPER·

WM·PENSTONE·

"LATE WROVGHT-
IRON WORK·"BY·

ARCHITECT~

D

PLATE VI.

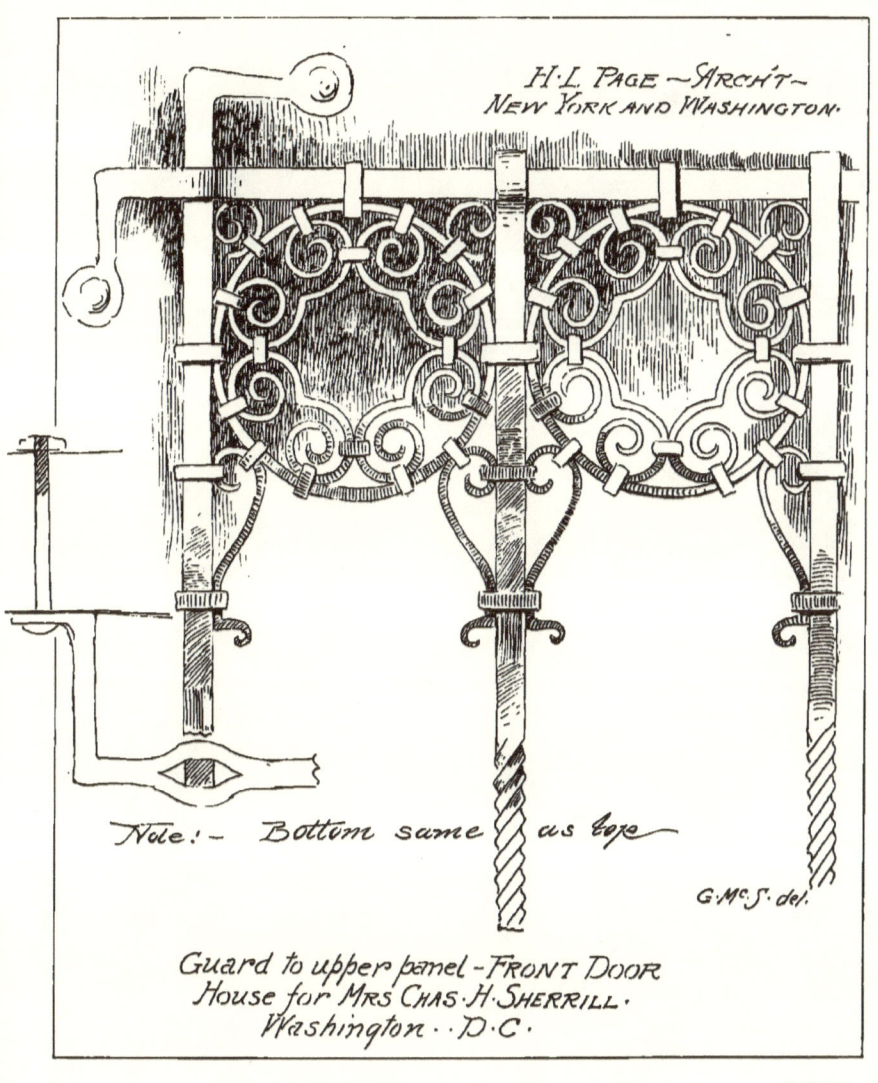

H·L· PAGE ~ARCH'T~
NEW YORK AND WASHINGTON·

Note!~ Bottom same as top

G·Mc·S·del.

Guard to upper panel - FRONT DOOR
House for MRS CHAS·H·SHERRILL·
Washington··D·C·

PLATE VII.

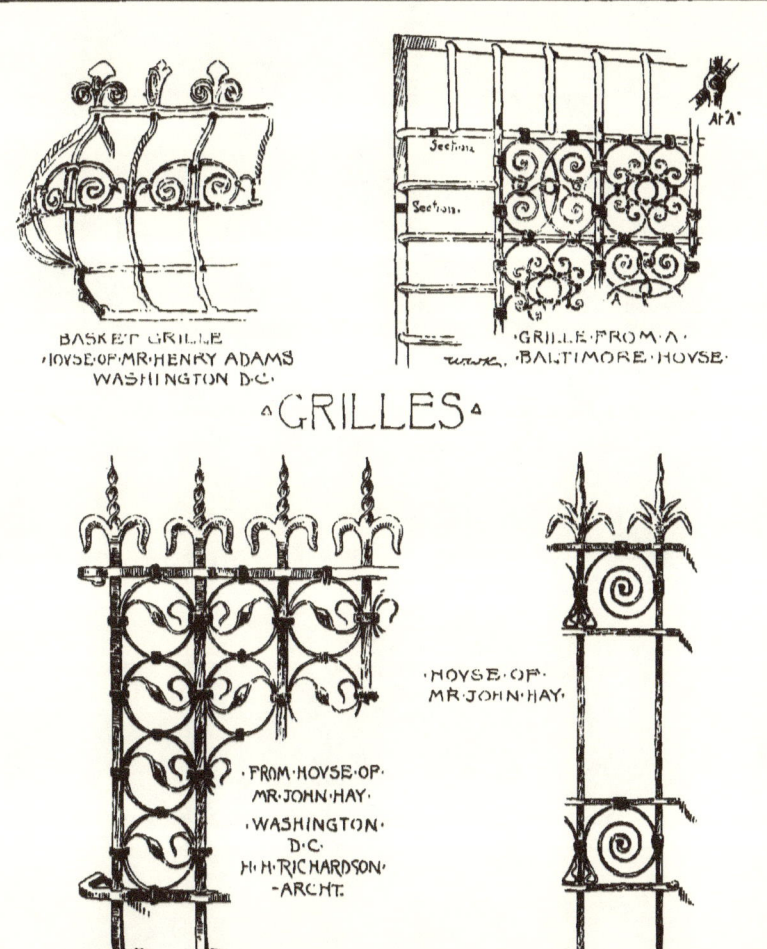

BASKET GRILLE
HOVSE OF MR HENRY ADAMS
WASHINGTON D.C.

GRILLE FROM A
BALTIMORE HOVSE

At "A"

Section

Section

^GRILLES^

FROM HOVSE OF
MR JOHN HAY
WASHINGTON
D.C.
H.H. RICHARDSON
ARCHT.

HOVSE OF
MR JOHN HAY

MESSRS KRVG AND SON BALTIMORE MD.

PLATE VIII.

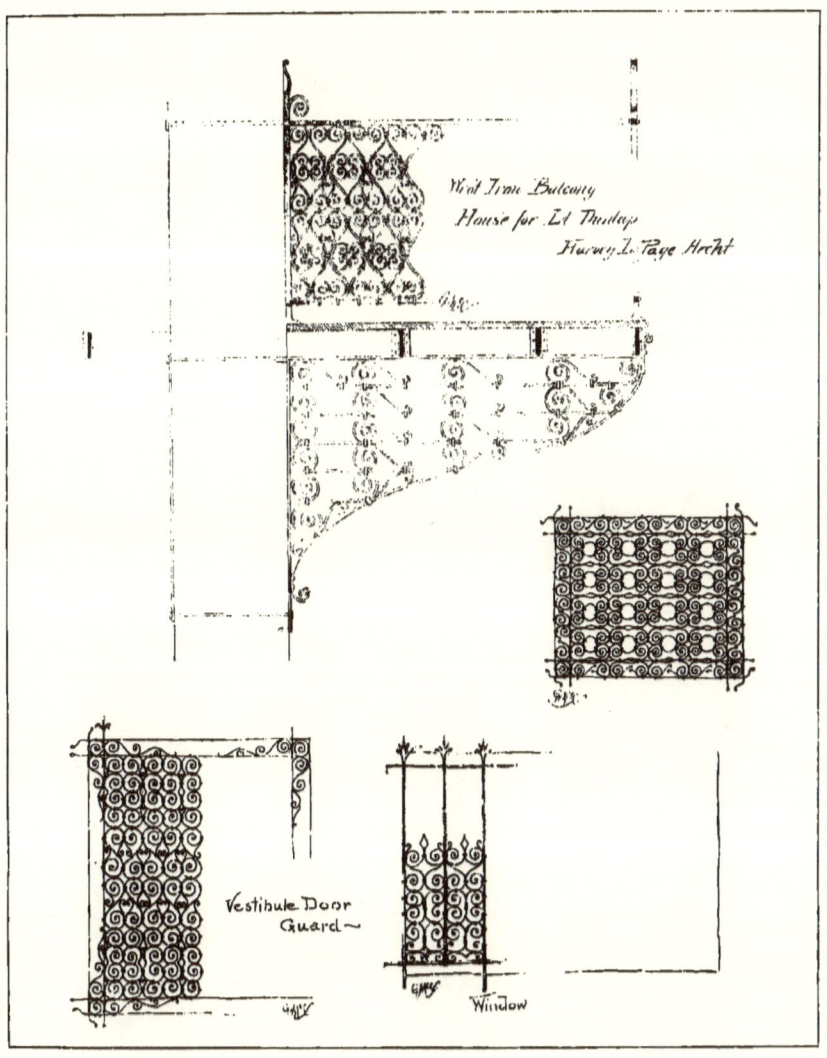

Wro't Iron Balcony
House for Lt Dunlap
Harvey L. Page Archt

Vestibule Door
Guard~

Window

PLATE IX.

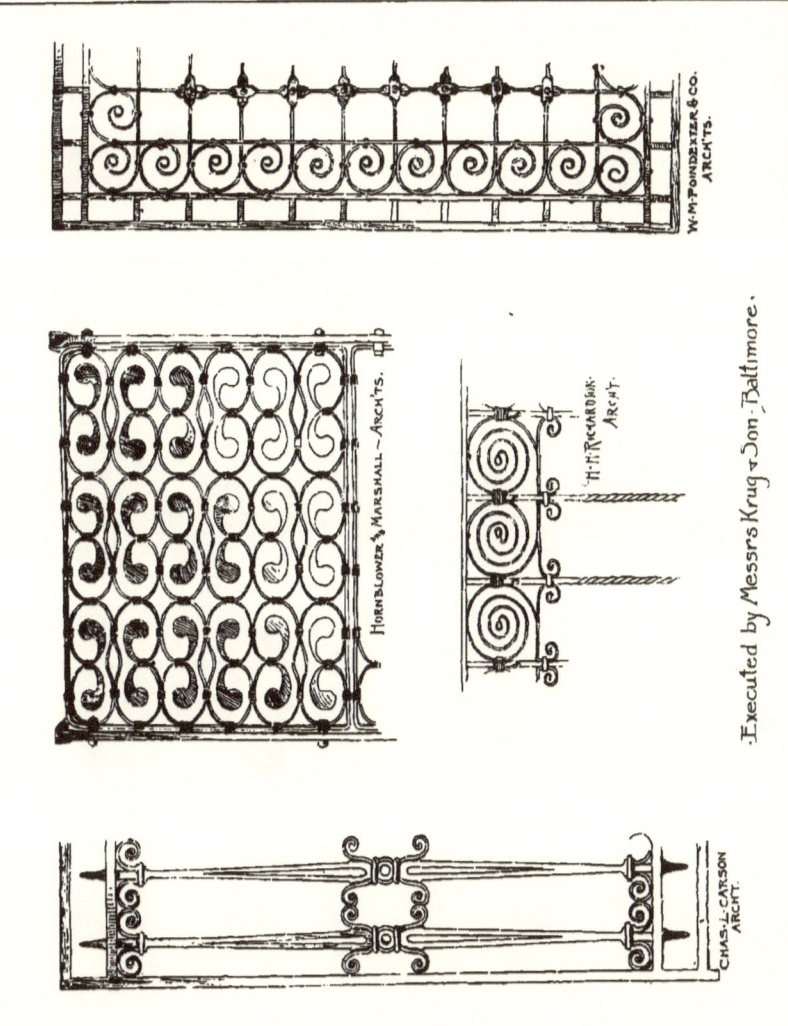

W. M. POINDEXTER & CO. ARCH'TS.

HORNBLOWER & MARSHALL—ARCH'TS.

H. R. RICHARDSON. ARCHT.

Executed by Messrs Krug & Son, Baltimore.

CHAS. L. CARSON ARCHT.

PLATE X.

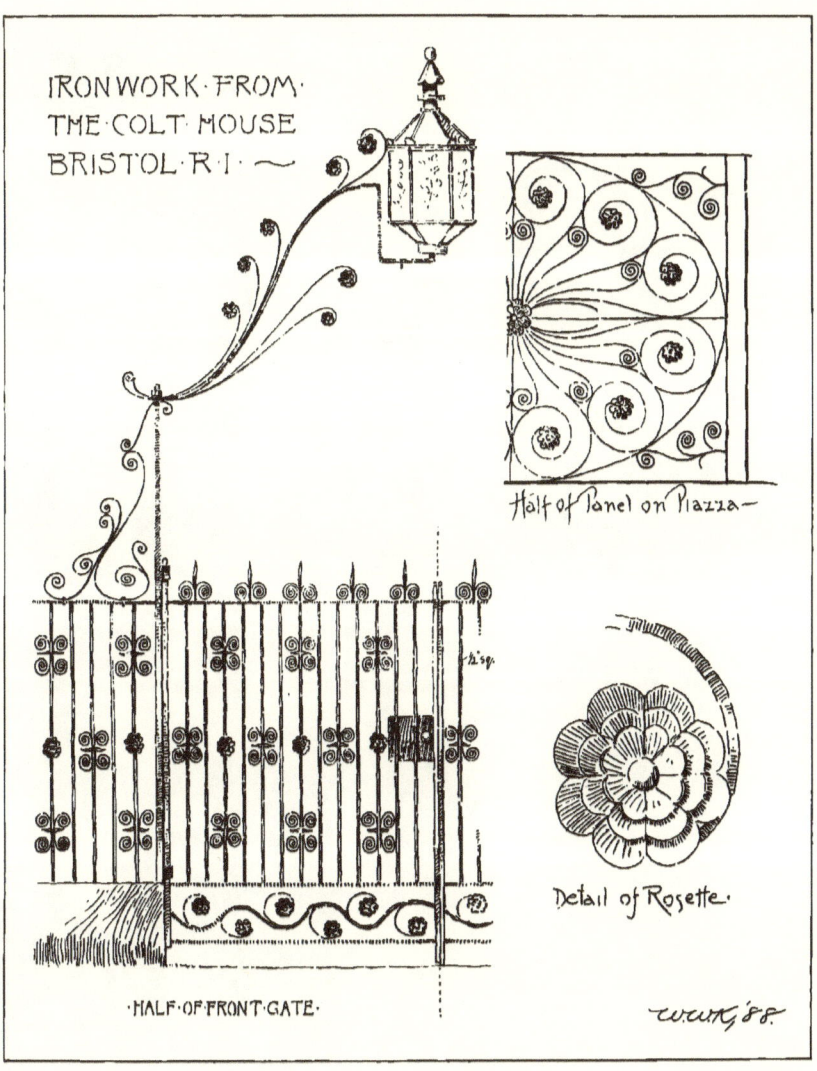

IRONWORK·FROM·
THE·COLT·HOUSE
BRISTOL·R·I·~

Half of Panel on Piazza—

Detail of Rosette.

·HALF·OF·FRONT·GATE·

WWK'88.

PLATE XI.

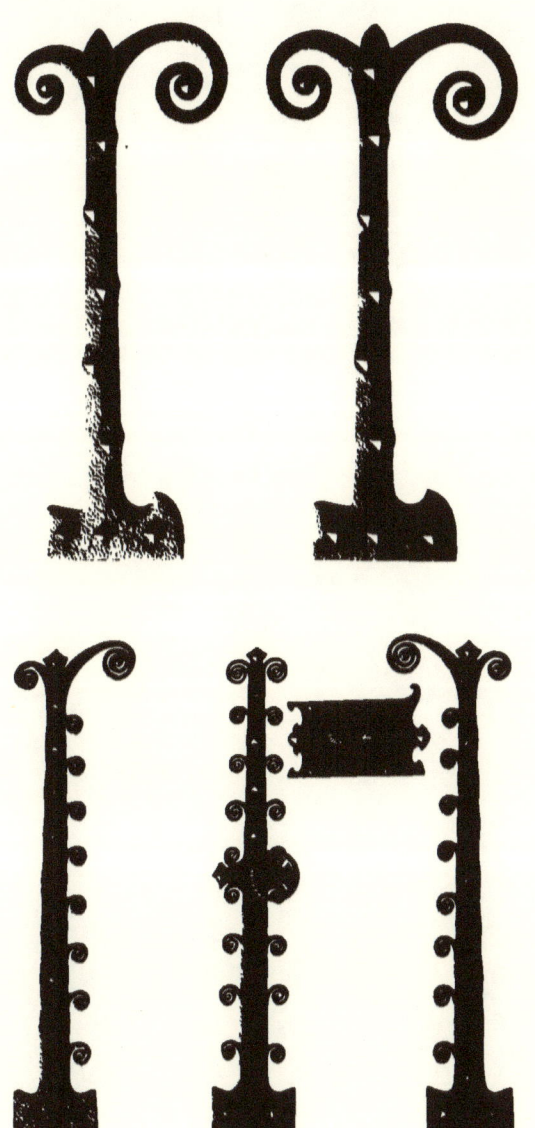

From Front Door of House of Mr Henry Adams
Washington D C

From Front Door of House of Mr John Hay
Washington D C

MR H H RICHARDSON
ARCHITECT

PLATE XII

PIACENZA
ITALY

SOUTH
KENSINGTON.

GERMAN
HINGE

PLATE XIII.

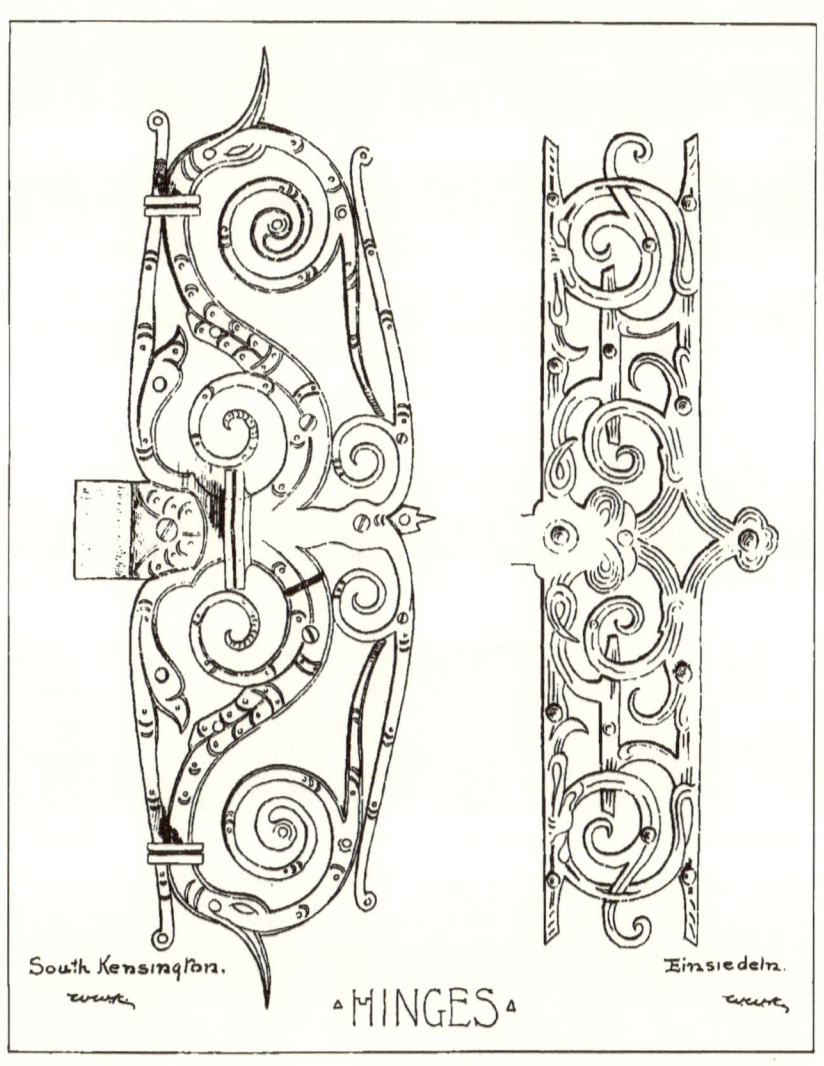

South Kensington.

Einsiedeln.

⁂HINGES⁂

PLATE XIV.

Rathaus, Lucerne.

LUCERNE

·SWISS·HINGES·

17th Century.
Museum at
Zurich.

PLATE XV.

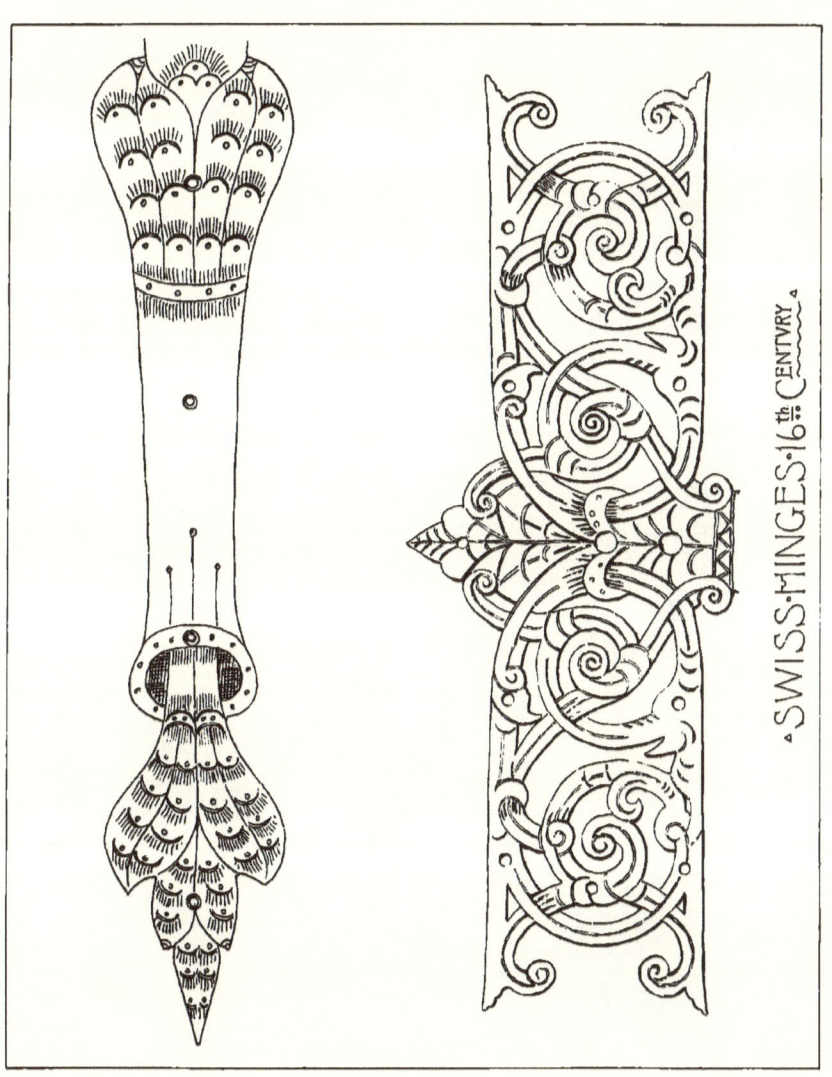

·SWISS·MINGES·16ᵗʰ CENTVRY·

PLATE XVI.

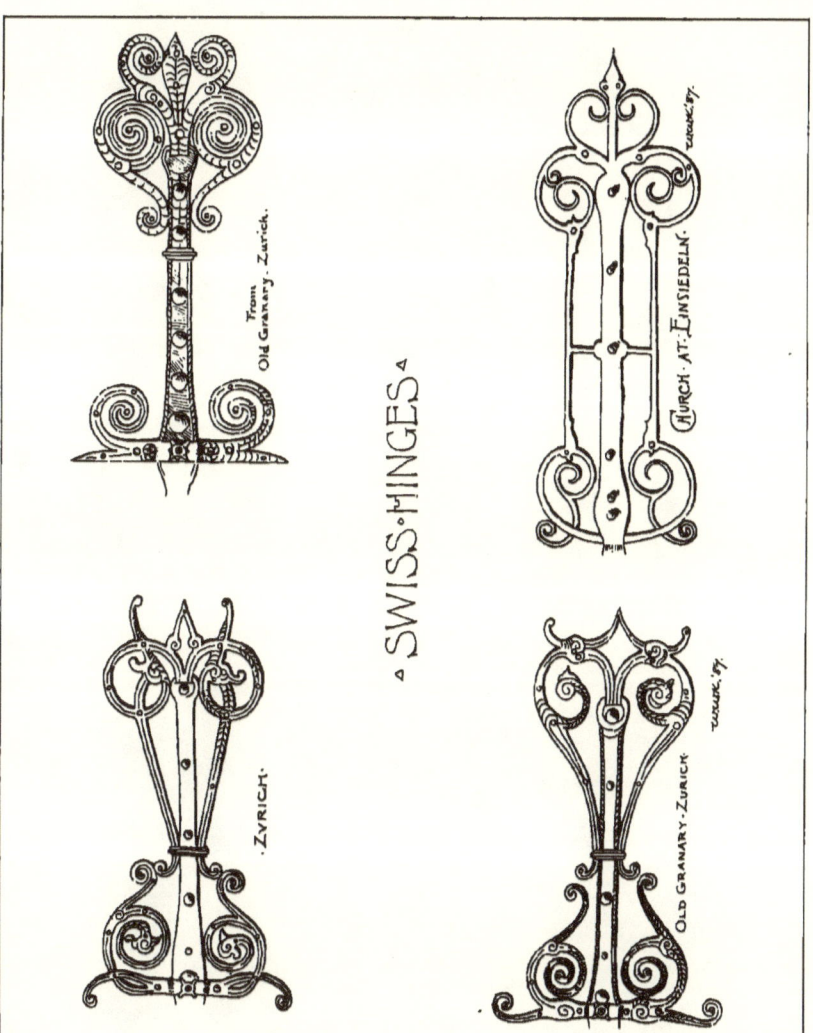

From
Old Granary . Zurich.

Church . at . Einsiedeln .

. Swiss . Minges .

. Zurich .

Old Granary . Zurich .

PLATE XVII.

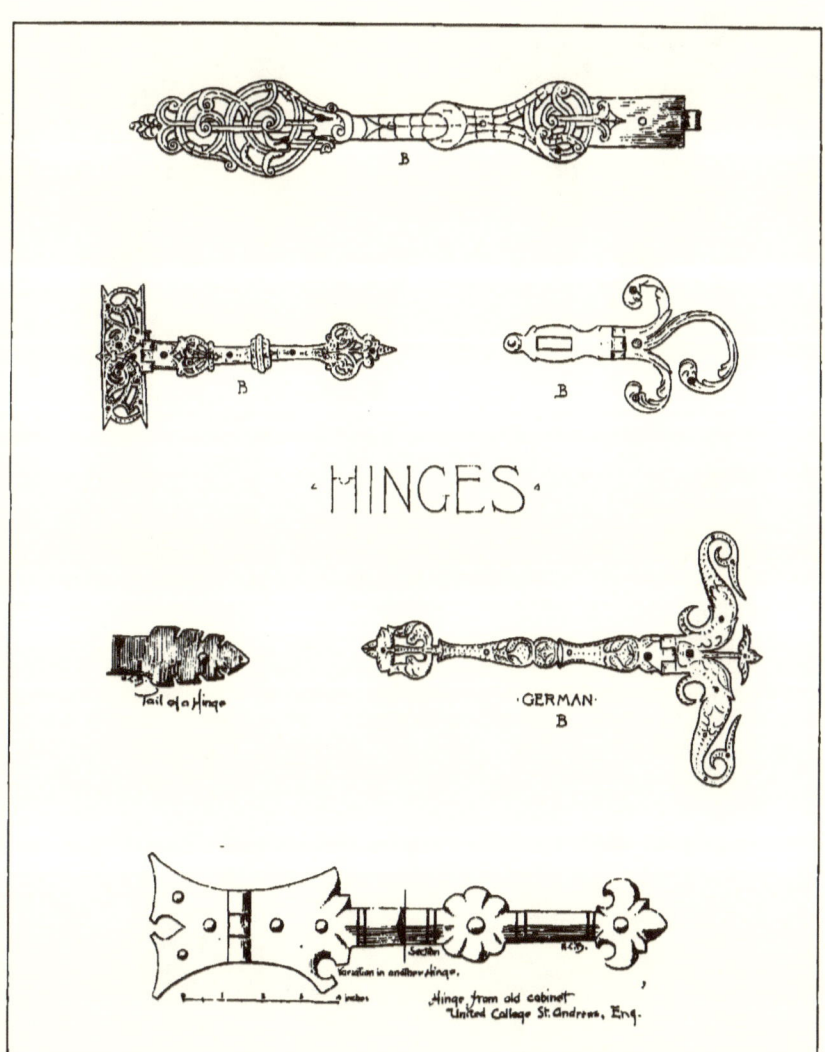

·HINGES·

Tail of a Hinge

·GERMAN·
B

Variation in another Hinge.

Section

Hinge from old cabinet
United College St. Andrews, Eng.

PLATE XVIII.

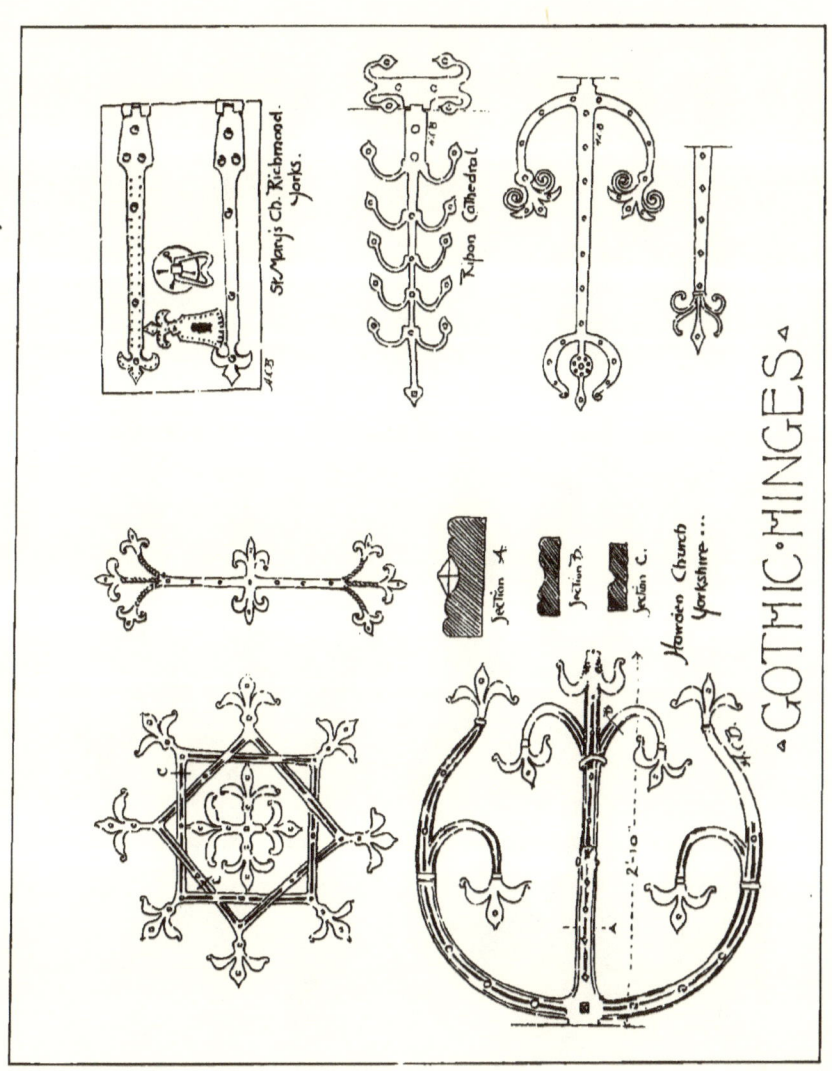

St Mary's Ch. Richmond. Yorks.

Ripon Cathedral.

Section A.

Section B.

Section C.

Howden Church Yorkshire...

GOTHIC HINGES

PLATE XIX.

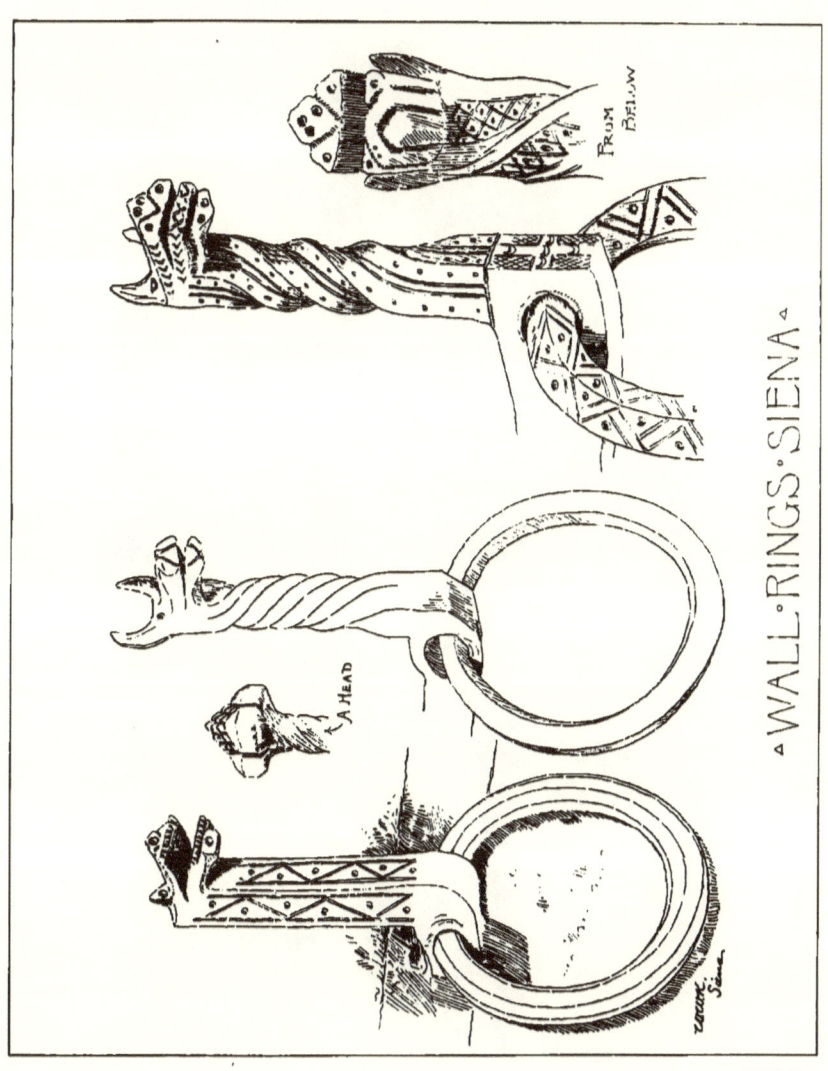

FROM BELOW

A HEAD

⊲ WALL · RINGS · SIENA ⊳

PLATE XX.

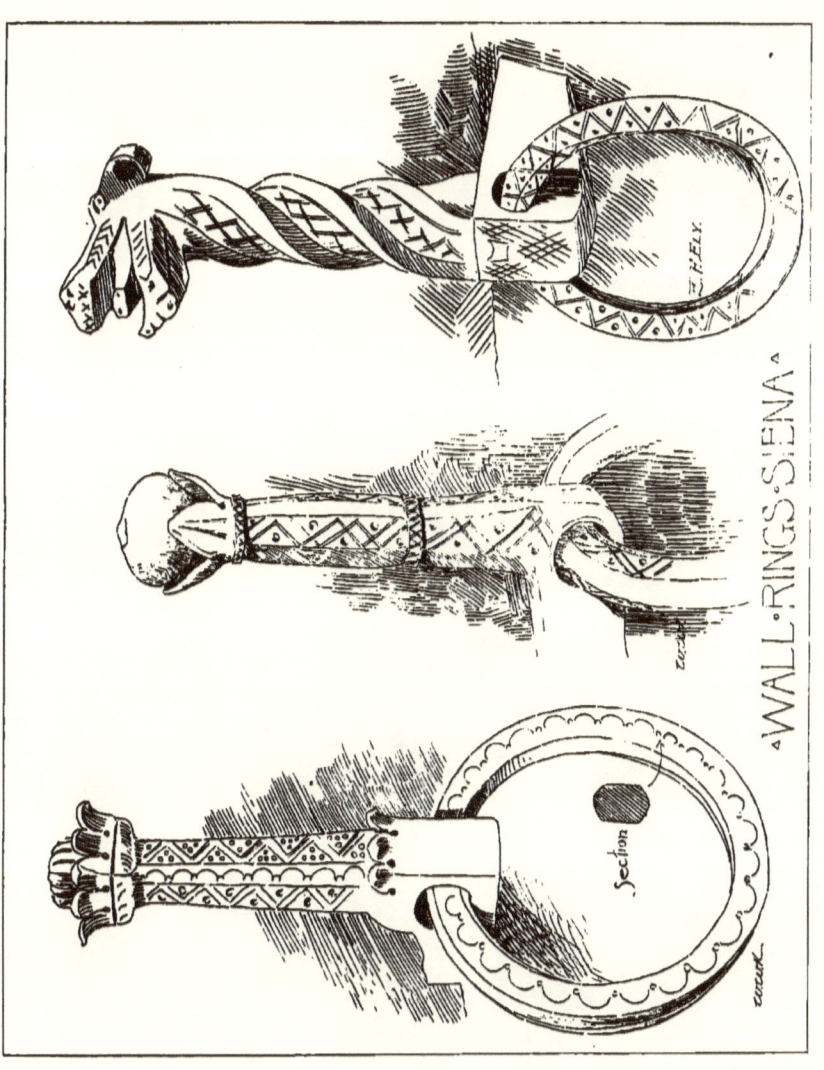

WALL·RINGS·SIENA·

PLATE XXI.

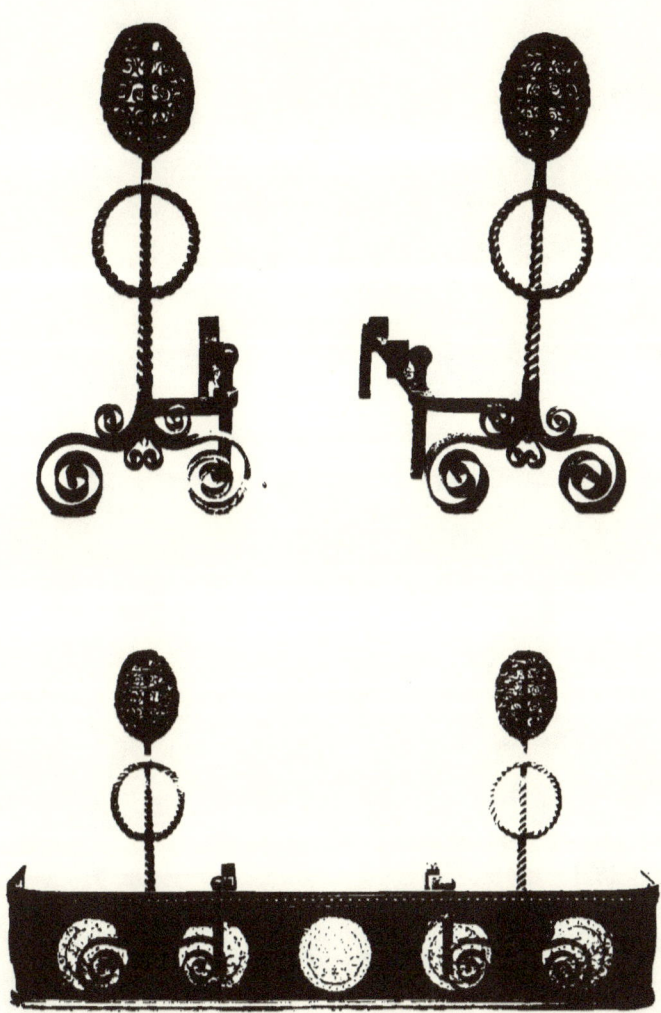

From House of Mr John Hay.
Washington, D C.

Plate XXII

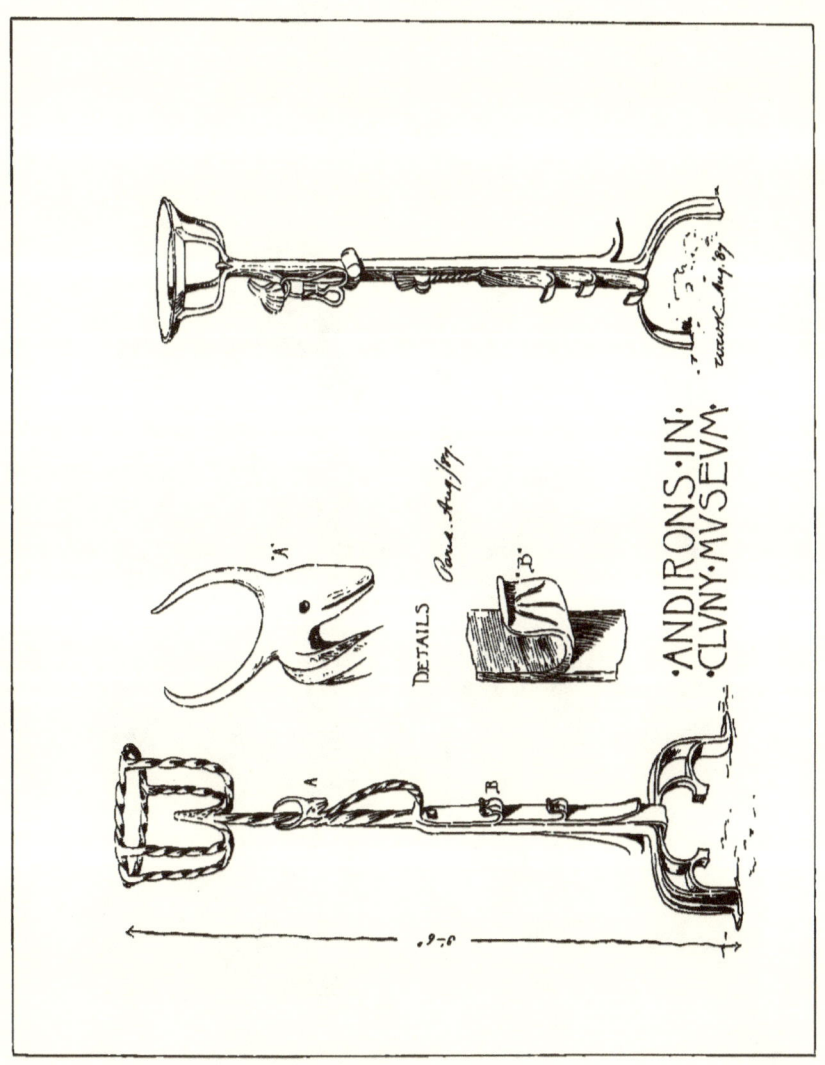

DETAILS

·ANDIRONS·IN·
·CLVNY·MVSEVM·

PLATE XXIII.

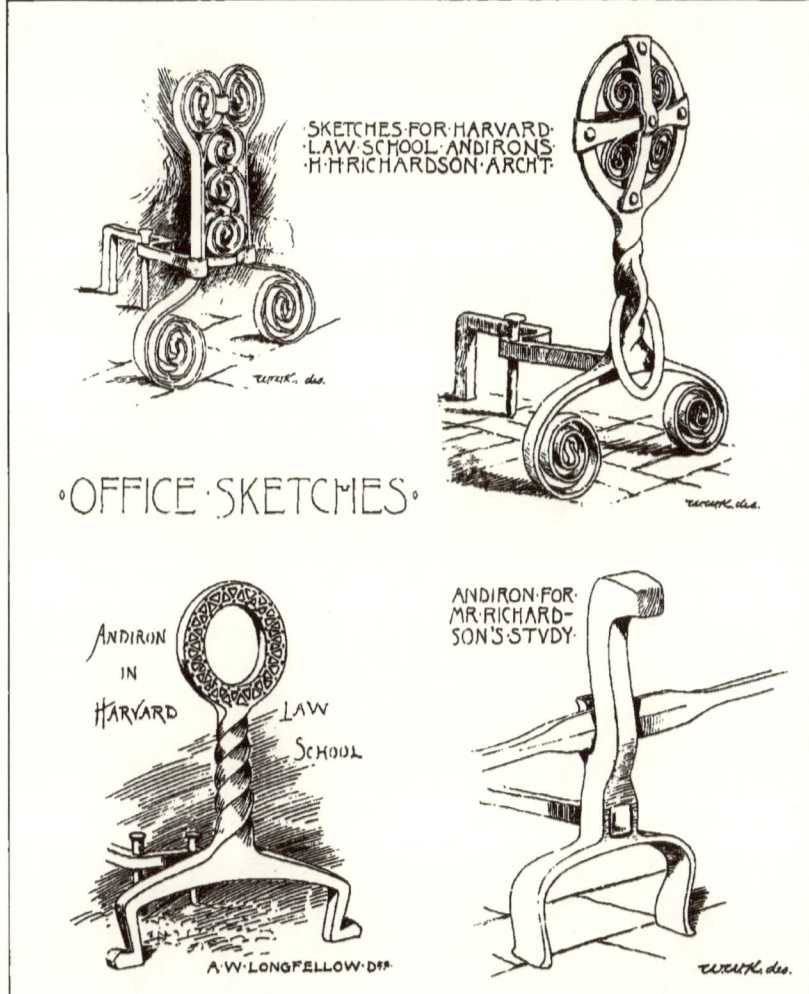

SKETCHES·FOR·HARVARD·
·LAW·SCHOOL·ANDIRONS·
·H·H·RICHARDSON·ARCH'T·

·OFFICE·SKETCHES·

ANDIRON IN HARVARD LAW SCHOOL

A·W·LONGFELLOW·Del.

ANDIRON·FOR·
MR·RICHARD-
SON'S·STVDY·

PLATE XXIV.

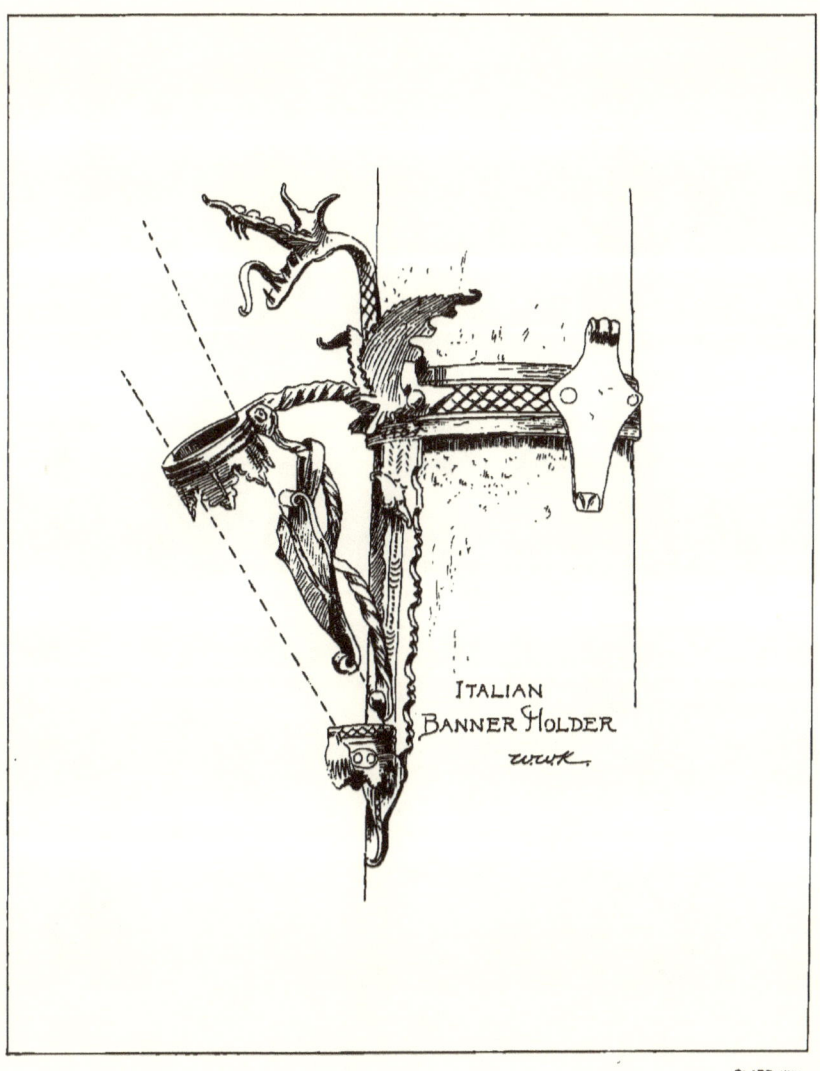

ITALIAN
BANNER HOLDER

PLATE XXV.

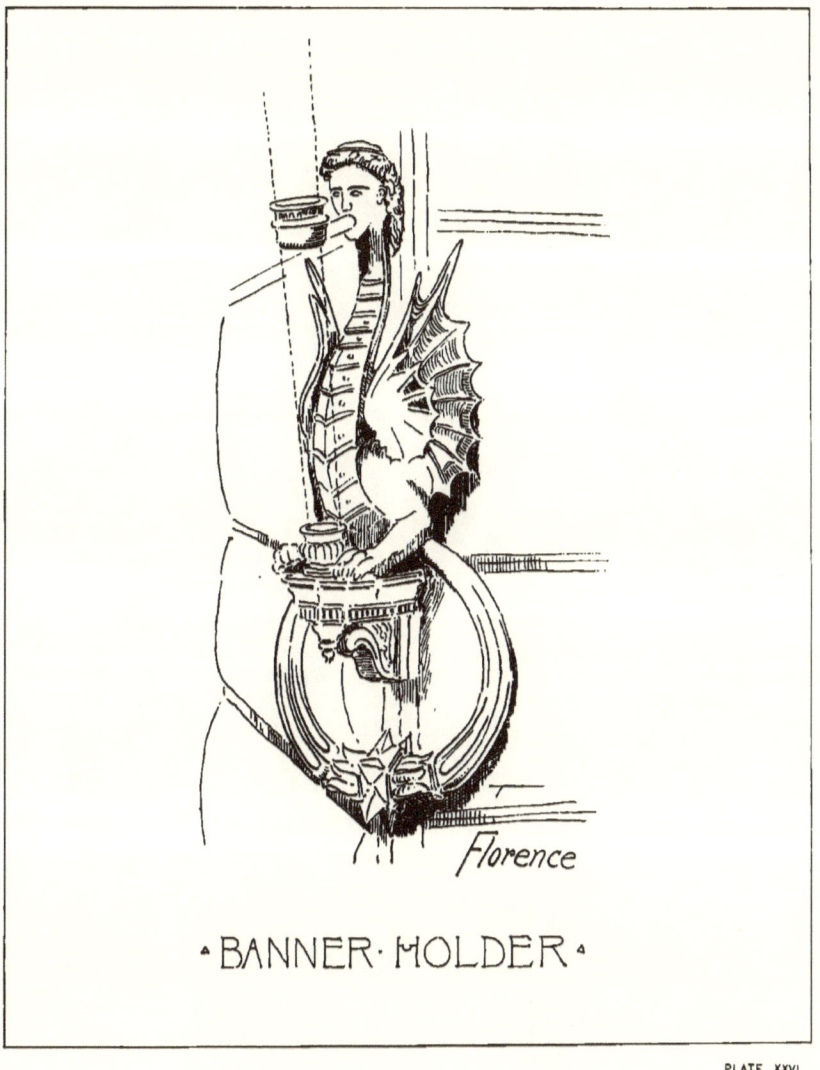

Florence

⋅ BANNER ⋅ HOLDER ⋅

PLATE XXVI.

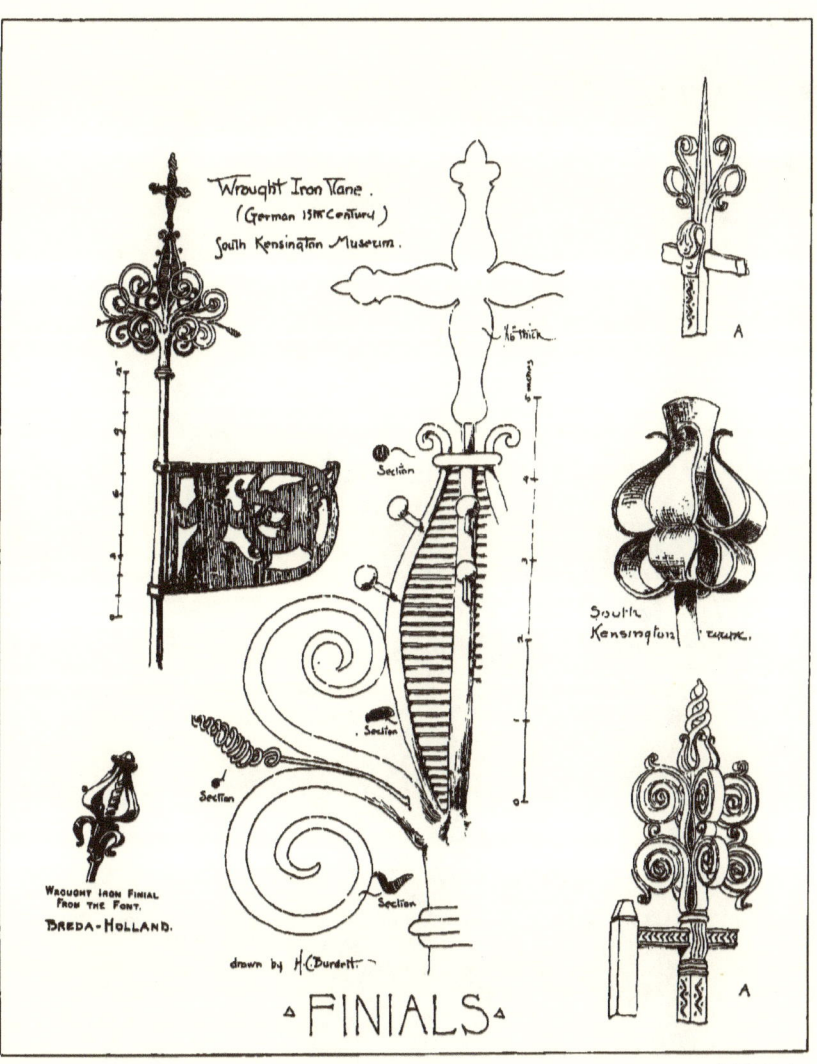

Wrought Iron Vane.
(German 15th Century)
South Kensington Museum.

A

South
Kensington

Wrought Iron Finial
From the Font.
BREDA-HOLLAND.

drawn by H.C.Burdett

⊿ FINIALS ⊿

A

PLATE XXVII.

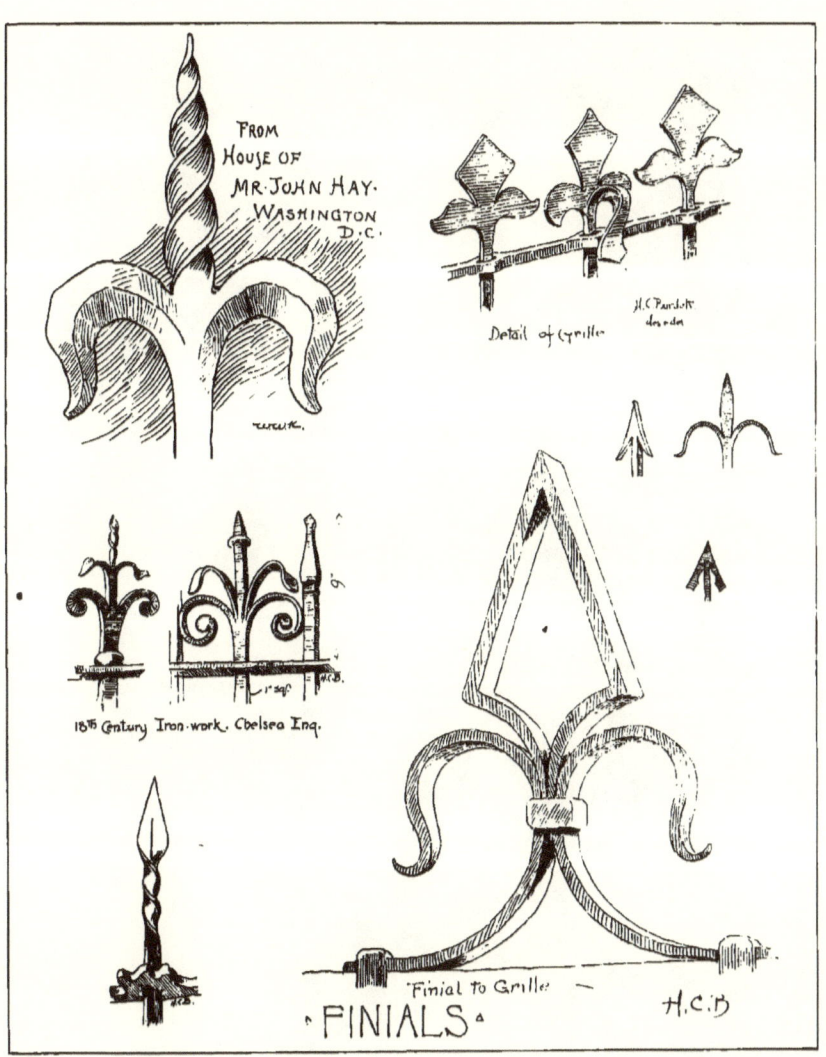

FROM
HOUSE OF
MR·JOHN·HAY·
WASHINGTON
D·C·

Detail of Grille

H.C.Purdott
des·n·den

18th Century Iron·work. Chelsea Inq.

'Finial to Grille —

·FINIALS·

H.C.D

PLATE XXVIII.

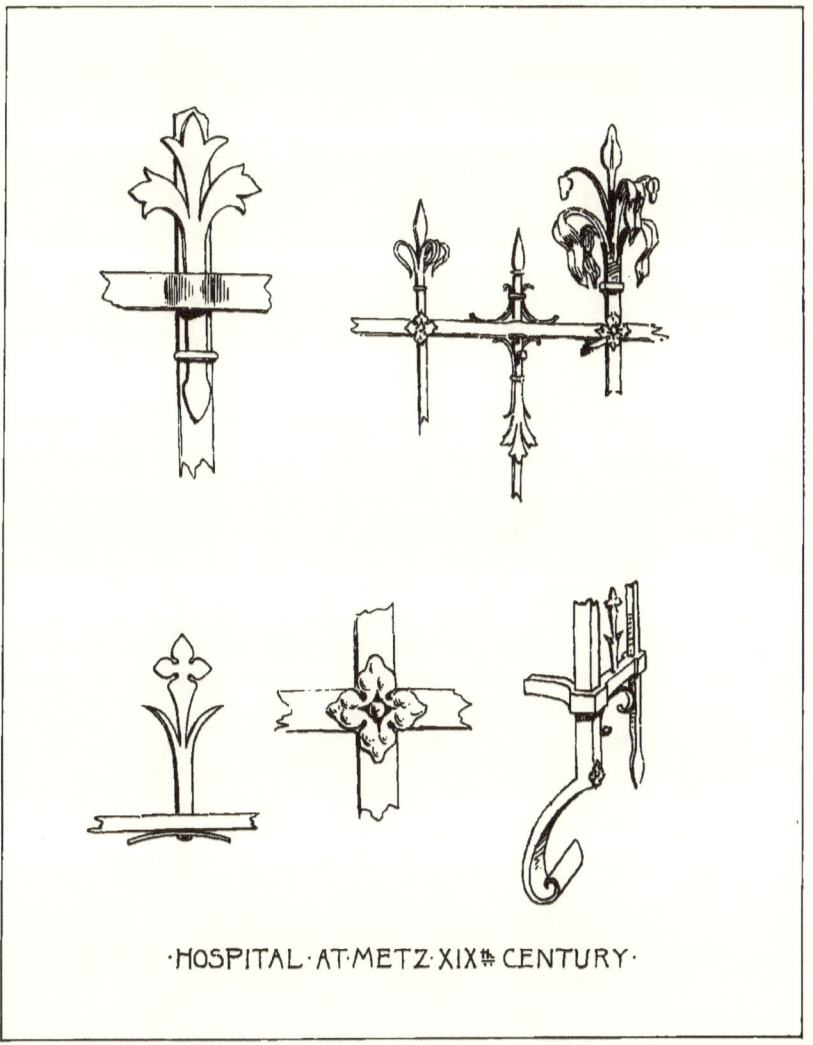

·HOSPITAL·AT·METZ·XIXth·CENTURY·

PLATE XXIX.

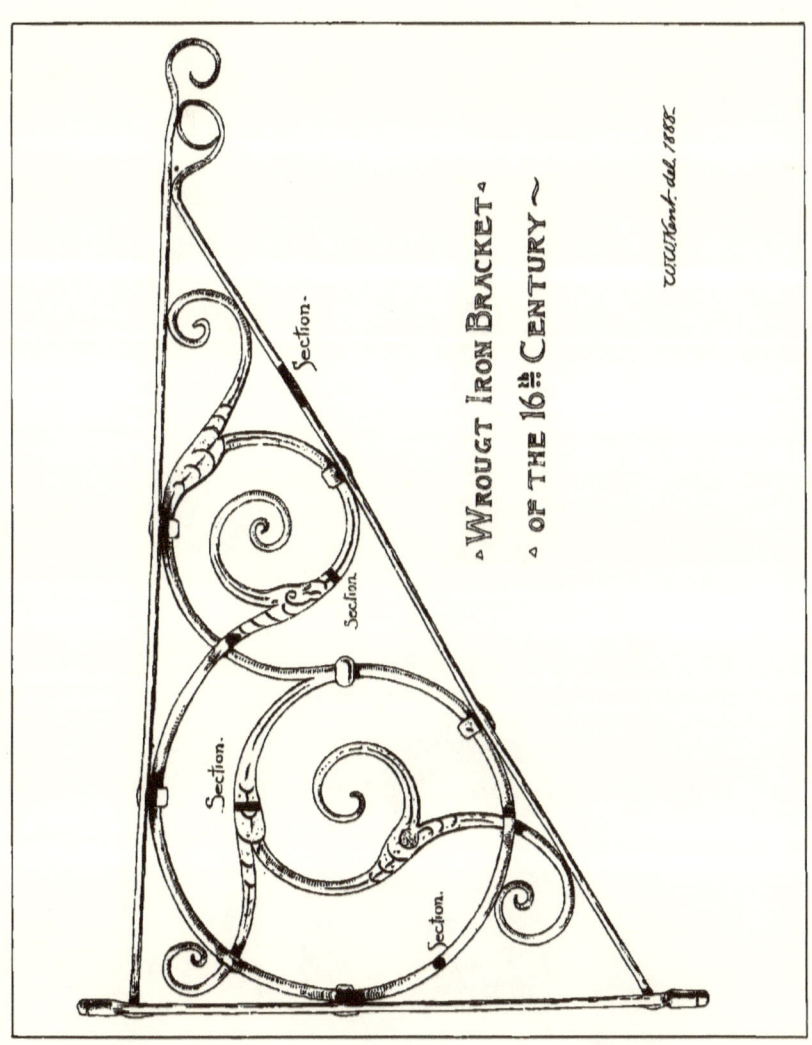

WROUGT IRON BRACKET
OF THE 16TH CENTURY

Section.
Section.
Section.
Section.

W.W.Kent. del. 1887.

PLATE XXX.

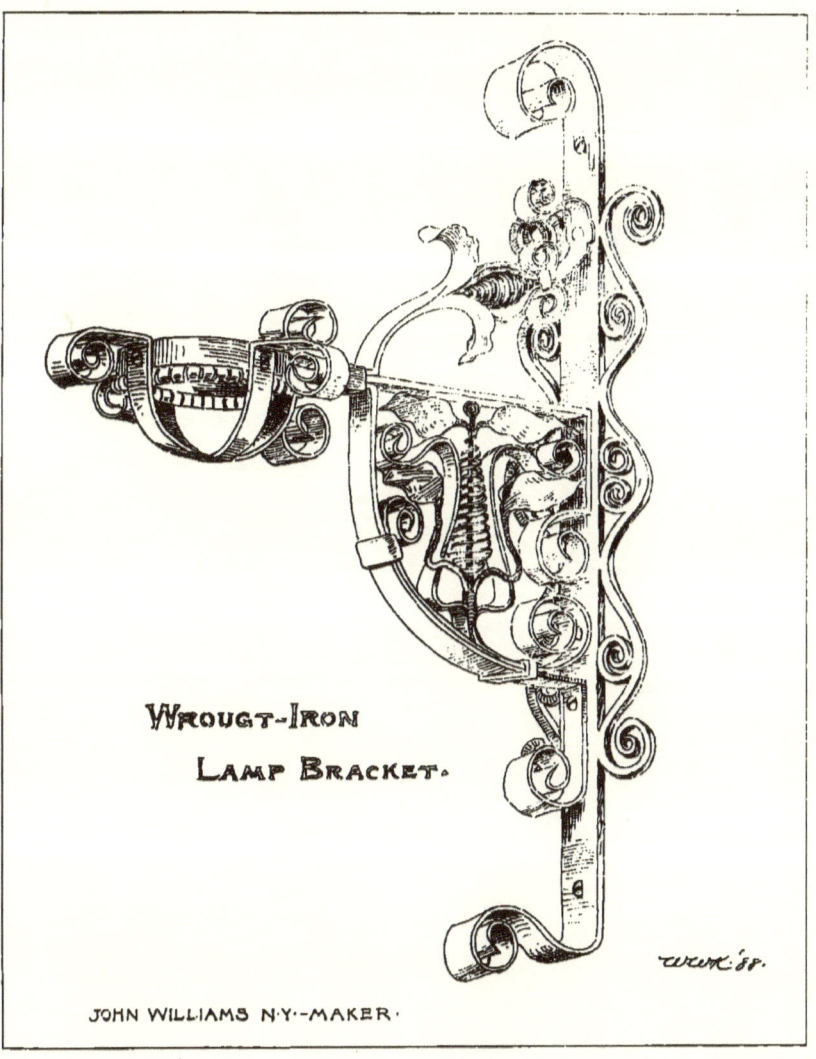

WROUGT-IRON
LAMP BRACKET.

wew '88.

JOHN WILLIAMS N·Y·-MAKER·

PLATE XXXI.

Shutter Door Handle.

Handle from Darnich Tower
Melrose · ½ full size.
A

B

A

·HANDLES·

A

B

A

Lexington.

Iver der.
Louvre
France.

Dieppe.

Edinburgh.   Dieppe.

C

PLATE XXXII.

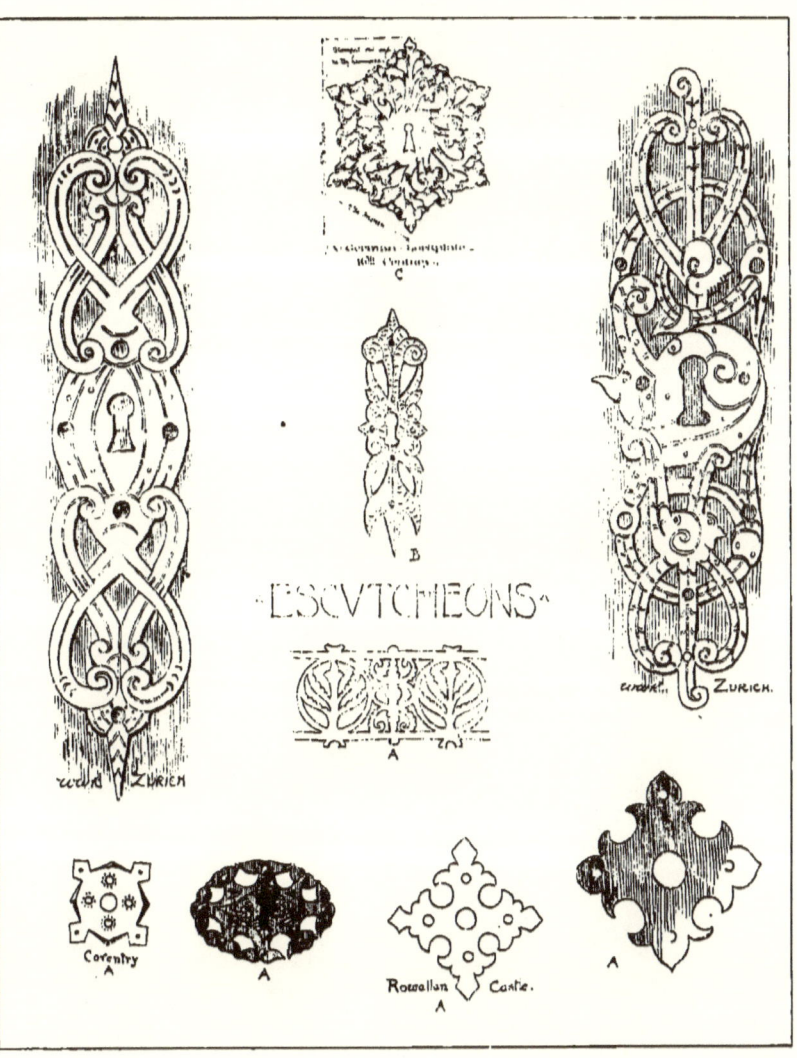

ESCVTCHEONS

C
A German Lockplate.
16th Century.

B

A

Zurich.

Zurich.

Coventry
A

A

Rowallan Castle.
A

A

PLATE XXXIII

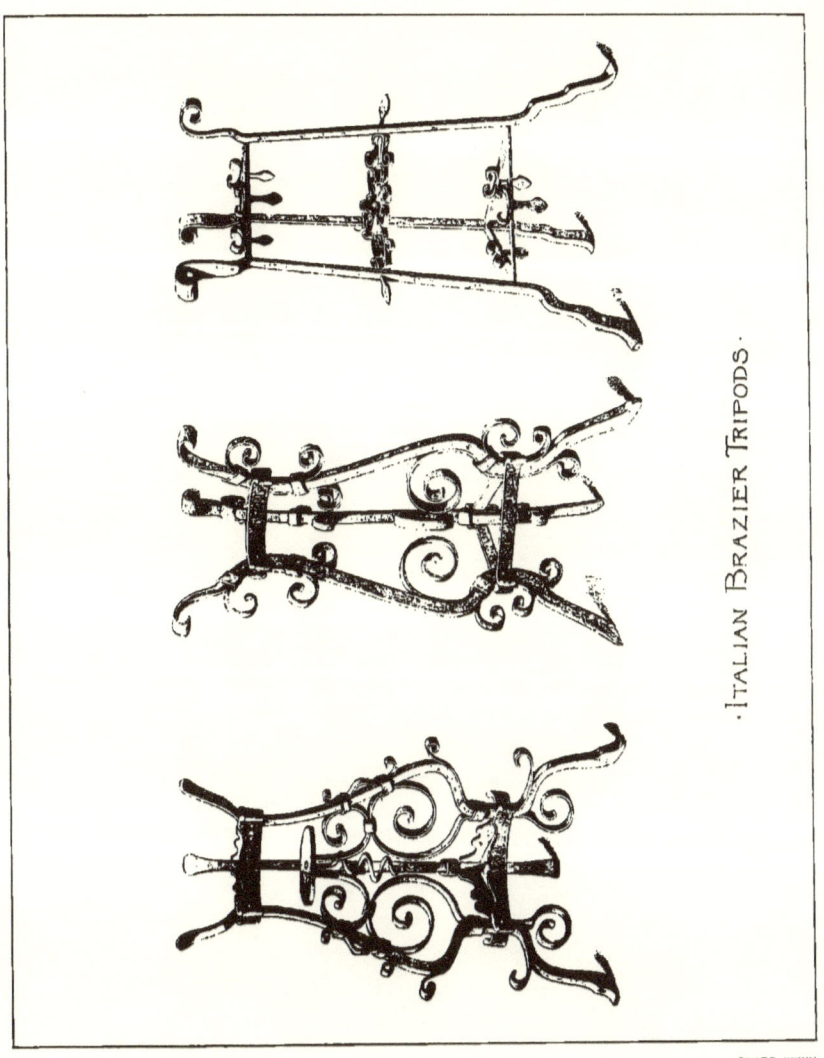

·ITALIAN BRAZIER TRIPODS·

PLATE XXXIV.

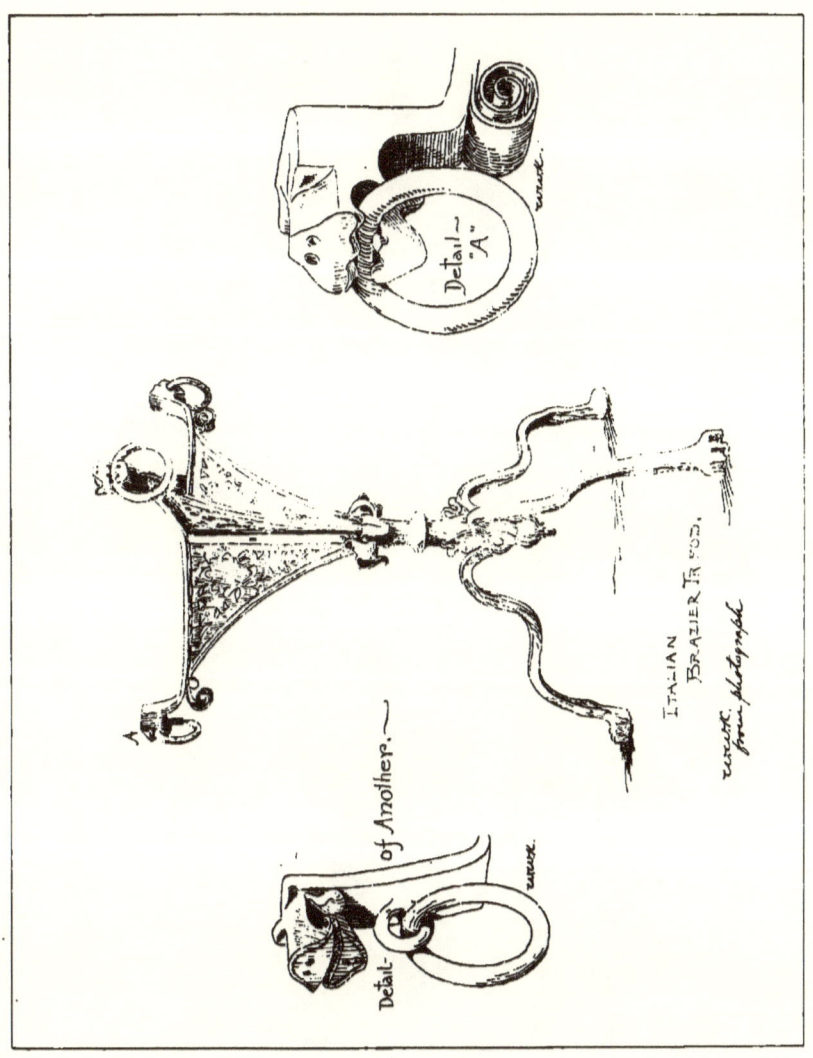

Detail— "A"

of Another.—

Detail—

ITALIAN
BRAZIER TRIPOD.

from photograph

PLATE XXXV.

www.ingramcontent.com/pod-product-compliance
Lightning Source LLC
Chambersburg PA
CBHW032154010726
47493CB00008BA/2697